Elasticity of Materials - Basic Principles and Design of Structures

Edited by Ezgi Günay

Published in London, United Kingdom

IntechOpen

Supporting open minds since 2005

Elasticity of Materials - Basic Principles and Design of Structures
http://dx.doi.org/10.5772/
Edited by Ezgi Günay

Contributors
Ezgi Günay, Pulkit Kumar, Moumita Mahanty, Amares Chattopadhyay, Duc-Nam Nguyen, Paul S. Nnamchi, Camillus S. Obayi, Katsufumi Hashimoto, Tomoki Shiotani, Takahiro Nishida and Nobuhiro Okude

First published in London, United Kingdom, 2019 by IntechOpen
IntechOpen is the global imprint of INTECHOPEN LIMITED, registered in England and Wales, registration number: 11086078, The Shard, 25th floor, 32 London Bridge Street
London, SE19SG – United Kingdom
Printed in Croatia

British Library Cataloguing-in-Publication Data
A catalogue record for this book is available from the British Library

Additional hard copies can be obtained from orders@intechopen.com

Elasticity of Materials - Basic Principles and Design of Structures
Edited by Ezgi Günay
p. cm.
Print ISBN 978-1-78984-927-1
Online ISBN 978-1-78984-928-8

For EU product safety concerns:
IN TECH d.o.o., Prolaz Marije Krucifikse Kozulić 3, 51000 Rijeka, Croatia, info@intechopen.com or visit our website at intechopen.com.

Meet the editor

Ezgi Günay has been an associate professor at the Mechanical Engineering Department at Gazi University in Turkey since 2010. She graduated from the Engineering Sciences Department (Applied Mechanics Department) of the Engineering Faculty at the Middle East Technical University (METU, Ankara, Turkey) in 1985. She received her Master of Science degree in 1989 from the same department. The title of the thesis was "Development of a Preprocessor and Modification of a Finite Element Procedure for the Analysis of Metal Forming Processes" (December 1989, METU). She graduated with her PhD degree at Gazi University from the Mechanical Engineering Department in 1996. The title of the thesis was "A Nonlocking Finite Element Model for Nonlinear Analysis of Thin and Thick Composite Plates." She worked as a research assistant between 1993 and 1999 and studied academically by giving basic courses as an assistant professor between 1999 and 2010 at the same department. During these years she has given courses in the following subjects: Technical Drawing, FORTRAN-Computer Programming Languages, Applied Mathematics for Mechanical Engineers, Differential Equations, Statics, Dynamics, Strength of Materials, Introduction to Numerical Analysis, Introduction to Composite Materials, Introduction to Finite Element Analysis (FEA), Finite Element Method (FEM), Plate and Shell Theories, and Elasticity. She has authored about 40 papers published in both national and international proceedings and journals. She has three chapters published in international books. She has written two programming e-books in FORTRAN language (co-author) for engineering students.

Ezgi Günay is presently working on the following subjects: finite element linear and geometrically nonlinear analyses of fiber composite-stiffened shell and plate structures, linear and nonlinear FEA buckling analysis of fiber composite-stiffened plates, numerical and experimental studies on micromechanical investigation of layered fiber composites and natural fiber composite wood, FEA and analytical studies on single-walled carbon nano-tubes (SWCN), FEA of stress transfer mechanisms from matrix to fiber in SWCN-reinforced nanocomposites, and experimental and FEA studies on natural fiber composite wood (transversely isotropic) materials.

Contents

Preface XI

Section 1
General Theorems in Elasticity 1

Chapter 1 3
Introductory Chapter: Analytical and Numerical Approaches in Engineering Elasticity
by Ezgi Günay

Chapter 2 11
An Overview of Stress-Strain Analysis for Elasticity Equations
by Pulkit Kumar, Moumita Mahanty and Amares Chattopadhyay

Section 2
Engineering Applications in Theory of Elasticity 31

Chapter 3 33
FEA and Experimentally Determination of Applied Elasticity Problem for Fabricating Aspheric Surfaces
by Duc-Nam Nguyen

Chapter 4 53
Concept of Phase Transition Based on Elastic Systematics
by Paul S. Nnamchi and Camillus S. Obayi

Chapter 5 73
Repair Inspection Technique Based on Elastic-Wave Tomography Applied for Deteriorated Concrete Structures
by Katsufumi Hashimoto, Tomoki Shiotani, Takahiro Nishida and Nobuhiro Okude

Preface

During the nineteenth century, mechanics was mainly considered to be a part of applied mathematics. At the beginning of the twentieth century, all topics under the subject of mechanics, "solid mechanics" and "fluid mechanics," were defined with an index notation, and by re-explanation of these issues according to their coordinate systems, continuum mechanics has been created. With the gathering of all subjects in a single title, a new area has been opened in macroscopic (large-scale) mechanics. The exact solutions to linear/nonlinear elasticity problems have been discovered by applying analytical, numerical, and experimental methods and techniques to the new generation of computers and experimental equipment. Numerical solutions such as finite differences and finite elements methods have been extensively used to approximate the exact results.

The purpose of writing this book has been summarized in three main concepts. The first objective was to provide the basic information and principles about the history and theory of elasticity. The second was to explain the fundamental equations. The third was to present the recent engineering application studies that have been collected under the following basic headings: "FEA and Experimental Determination of Applied Elasticity Problems for Fabricating Aspheric Surfaces," "Repair Inspection Technique based on Elastic-Wave Tomography Applied for Deteriorated Concrete Structures," and "Concept of Phase Transition Based on Elastic Systematics."

The main subject of this book is engineering elasticity and consists of five chapters in two sections. The title of the first section is "General Theorems in Elasticity" and the first chapter of this section belongs to the editor and is entitled "Analytical and Numerical Approaches in Engineering Elasticity." In this chapter, the historical development of "elasticity theory" is presented briefly, and recent studies performed regarding the elasticity concept are categorized and listed according to their basic engineering problem groups. A literature survey has been performed and categorized between the years 2014 and 2018 and represented in a statistical plot.

The second chapter in the first section is entitled "A General Overview of Stress-Strain Analysis for the Elasticity Equations" and explains the results of elasticity equations and the analysis of stress, strain, and stress–strain relationships through particular sections. In this section the concept of normal and shear stresses, principal stresses, plane stress, Mohr's circle, stress invariants, stress equilibrium equations, linear elasticity, generalized Hooke's law, and stress–strain relationships for triclinic, monoclinic, orthotropic, transversely isotropic, fibre-reinforced, and isotropic materials are discussed by researchers P. Kumar, M. Mahanty, and A. Chattopadhyay.

The title of the second section is "Engineering Applications in Theory of Elasticity" and consists of three chapters on engineering applications on elasticity.

The first chapter of the second section is entitled "FEA and Experimental Determination of Applied Elasticity Problems for Fabricating Aspheric Surfaces" and is written by Dr. D.N. Nguyen. In this chapter, the elastic deformation machining method is explained in two cases "Elastic deformation machining method with mold" and "Elastic deformation machining method without mold." When the vacuum pressure was used in the construction of complex aspherical surfaces, the differential equations of an appropriate plate theory were solved and the amount of deviation of the circular plate was determined and the test results were presented comparatively. Finite element analysis results for "Elastic deformation machining process with mold" were presented by curves. The finite element model was designed for establishing the spherical surface through a simulation of contact processing between workpiece and mold surface. In conclusion it was clarified that the experimental results agreed greatly with FEA results.

The second chapter is entitled "Concept of Phase Transition Based on Elastic Systematics" and is presented by Dr. P.S. Nnamchi and Dr. C.S. Obayi. In this chapter, the authors present the actual scaling of phase transition-driven considerations, such as martensitic transformation and transformable shape memory formation via elastic constant systematics in terms of continuum mechanics. According to this chapter, the results of the scaling procedure and acoustic anisotropy with respect to the mechanical stability criteria of the polycrystals based on the elastic modulus are compatible with the new experimental data obtained from the literature.

The third chapter is entitled "Repair Inspection Technique Based on Elastic-Wave Tomography Applied for Deteriorated Concrete Structures" and was written by Dr. K. Hashimoto, Dr. T. Shiotani, Dr. T. Nishida, and Dr. N. Okude. In this research, the testing results based on the internal damage assessment for the repair condition by applying elastic wave tomography and acoustic emission tomography are presented from a concrete pier, concrete wall, and slab obtained form. Determining the 3D velocity distribution, the repair effects of the epoxy injection method and the patch repair method are quantitatively evaluated and results are explained.

The first section of this book includes the recently published literature on elasticity concepts and basic theoretical knowledge. In the second section, researchers have focused on the engineering applications and on the use of elasticity theory by experimental, numerical, and analytical studies.

Acknowledgments

The editor would like to express her great appreciation to Ms. Romina Skomersic and Opr. Dr. Ayşe Günay for their valuable supporting studies regarding the formation of this scientific book. My special thanks are extended to the staff of IntechOpen for publishing this book.

Assoc. Prof. Dr. Ezgi Günay
Gazi University,
Engineering Faculty,
Mechanical Engineering Department,
Ankara, Turkey

Section 1

General Theorems in Elasticity

Chapter 1

Introductory Chapter: Analytical and Numerical Approaches in Engineering Elasticity

Ezgi Günay

1. Introduction

In this section, the historical development of the "elasticity theory" was presented briefly, and recent studies performed about the elasticity concept were categorized and listed according to their basic engineering problem groups. The mentioned literature survey has been performed by searching the keywords "elasticity," "analytic," and "solution" between the years 2014 and 2018. The most important general aspects of the "elasticity theory" were described in four groups as "the unknowns," "the used equations," "the modeling procedures," and "solution methods." In the future, in the consideration of these explained theoretical, numerical, and experimental properties, the researchers can be concentrating on the origin of the problem and new solution methods in deciding the exact nature of the material.

The elasticity concept of solid materials is the deformation with the external force application and recovery to its original shape after the forces removed. In the strength measurement of the material, stress (force per area) and strain (deformation per unit length) criteria have been used. The elasticity theory was presented in order to explain the basic theoretical concepts and their analytical solution methods, the deformations that were assumed to be very small and corresponding stress distributions. The classical elasticity theory was explained by theorems of "uniqueness of solution" and "existence of solution" as they have been declared in the basic mathematical concepts. The "uniqueness of solution" theorem was restricted to a single solution space by satisfying the related boundary or the initial conditions. If there were no any boundary or initial conditions, the solution space would have to be infinity. The "existence of solution" theorem was created by explaining the default displacement functions, checking the equilibrium equations for stress definitions, and satisfying the partial differential equations with the infrastructure of the default solutions. The purpose of the elasticity theory was the determination of this unique and exact solution in elastic region of the material. In linear elastic region, superposition method and combined loading applications are widely used in engineering.

2. Historical development in elasticity

The historical development of the concept of "elasticity" by considering mathematics, physics, and engineering mechanics was summarized in **Figure 1** [1, 2]. The scientific studies performed on engineering problems have been grouped as analytical, numerical, and experimental. The main solution techniques listed below

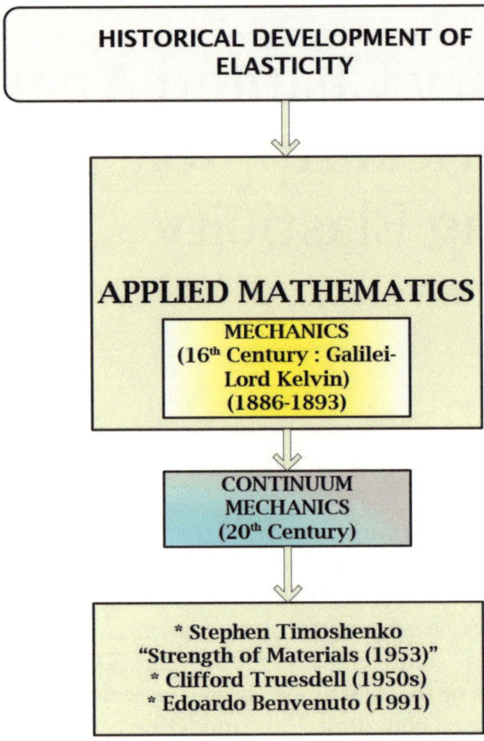

Figure 1.
Development of elasticity between the sixteenth and twentieth centuries.

form the first step aspects in performing the experiments and obtaining the numerical solutions by considering innovations: (i) characteristics of the solution methods, (ii) learning the mathematical theories, (iii) the physics of the problem, and (iv) learning the problem-solving methodologies. The second step aspects have been listed as "solving problems by mathematical techniques" and "obtaining new formulas." The scientific progress has been continued thanks to the studies done since the sixteenth century. The development in scientific area occurred in the elasticity concept has been summarized and visualized in the consideration of the scientists who have lived between the sixteenth and twentieth centuries and their studies [1, 2]. These famous scientists were Galilei (1564–1642), Mariotte (1620–1684), Hooke (1635–1703), Leibniz (1646–1716), Bernoulli (1700–1782), Baumgarten (1706–1757), Euler (1707–1783), Coulomb (1736–1806), Young (1773–1829), Poisson (1781–1840), Navier (1785–1836), Cauchy (1789–1857), Saint-Venant (1797–1886), Borchardt (1817–1880), Rankine (1820–1872), Kirchhoff (1824–1887), Maxwell (1831–1879), Clebsch (1833–1872), Kohlrausch (1840–1910), Amagat (1841–1915), Voigt (1850–1919), Mallock (1851–1933), Lamme (1864–1924), Röntgen (1872,1919), Synge (1897–1995), and Everett (1930–1982) (**Figure 1**).

3. Classification of engineering problems in the context of elasticity

In this section, the results of the literature review on elasticity were evaluated by referring to the articles (total number of articles, 157) between 2014 and 2018. Important information has gained from the literature survey about the elasticity theory and its related recent engineering solutions, as well as information about the theoretical, numerical, and experimental scientific researches and scientific innovations. The brief classification of the main engineering problems was summarized in **Figure 2**.

The studies evaluated in the literature review were listed below in 10 main headings. The distribution of articles corresponding to research concepts is presented in **Figure 3**. These are (1) historical development, (2) analytical and experimental studies related to the finite element method (FEM), (3) experimental studies, (4) analytical studies and finite element analysis (FEA), (5) analytical studies,

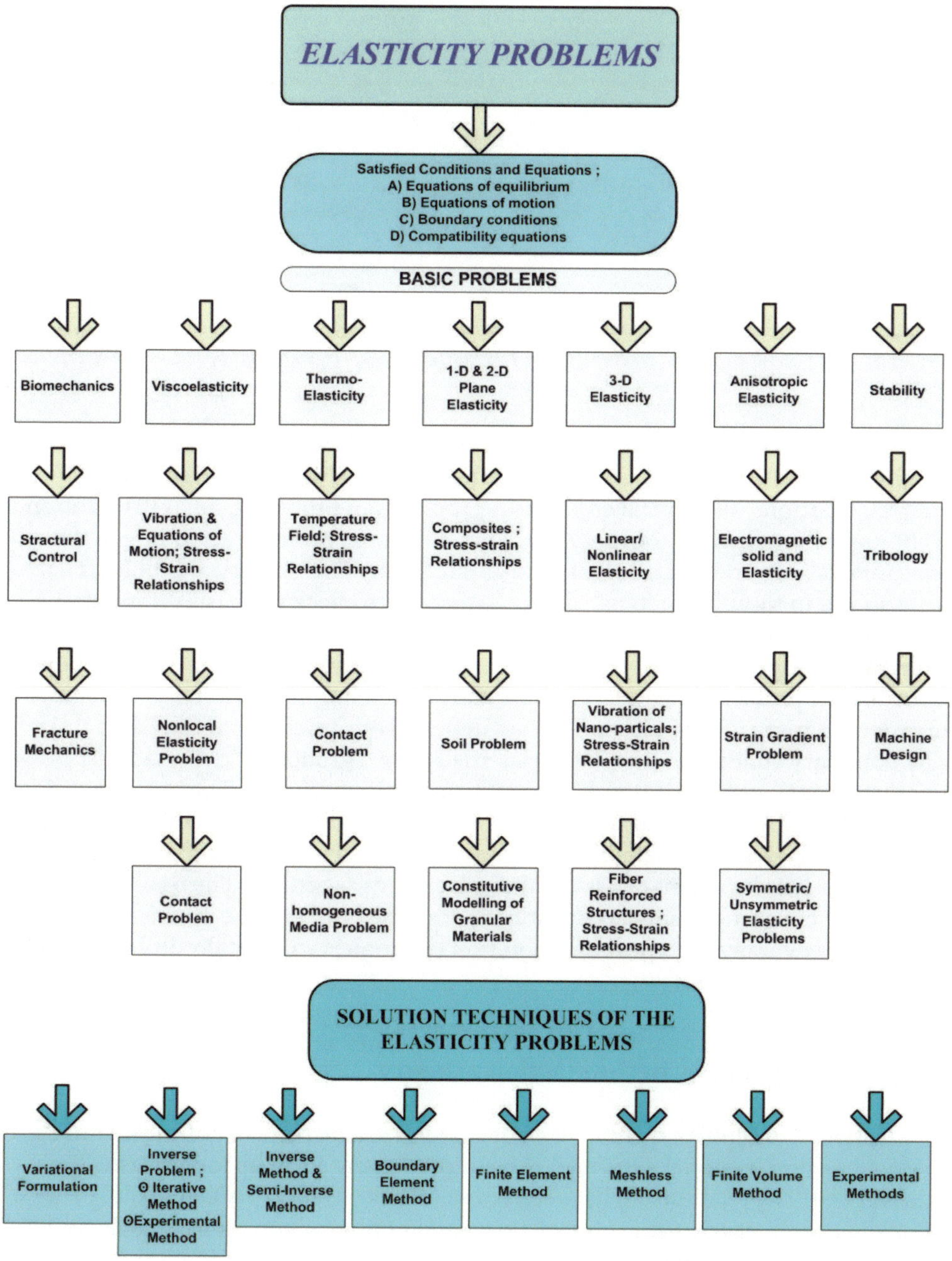

Figure 2.
Classification of the basic elasticity problems and their solution techniques.

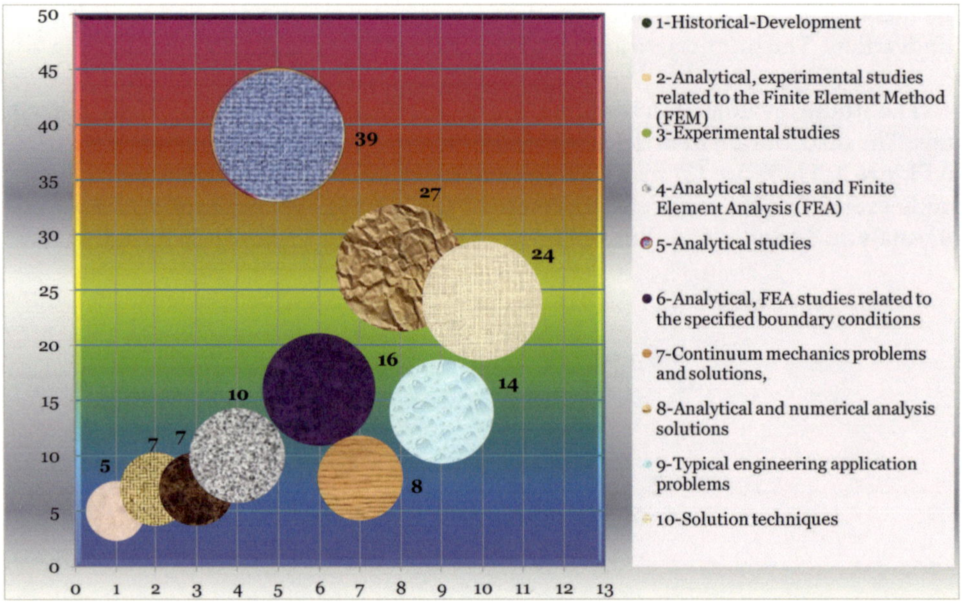

Figure 3.
The results of the literature review on elasticity were evaluated by referring to the 157 articles between 2014 and 2018.

(6) analytical and FEA studies related to the specified boundary conditions, (7) continuum mechanics problems and solutions, (8) analytical and numerical analysis solutions, (9) typical engineering application problems, and (10) solution techniques. The types of elasticity problems have been grouped according to the science innovations and related industrial applications. The numerical problems have been solved in three basic steps. The first step was to check the basic differential equations in terms of satisfaction with the placement of the estimated displacement functions. The second step was to check the "initial values" or the "boundary conditions" of the problem [3–5]. Values were substituted into the differential equations in order to satisfy the conditions at these defined coordinates or at time domains. The boundary conditions have been classified in two groups as "the essential" (displacement) and "the natural" (force) boundary conditions. The initial conditions were the first-stage variations named initiative and time-dependent variables. The third step was the satisfaction of the continuity conditions on the compatibility equations by means of assumed displacement functions. The basic elasticity problems were grouped into 26 subtitles as described in **Figure 2**. In this figure, the number of generally used proposed solution techniques analytically and numerically was equal to eight.

4. General principles in the elasticity theory

Elasticity concept is explainable by the natural elastic behavior of the materials. In elastic region, material deformed in a nonpermanent form up to the elastic limit was reached. The relationship between stress (σ) and strain (ε) under loading and unloading cases was explained by the linear and nonlinear equations. The slopes of the linear curves developed in linear elastic region were known as Young's modulus E, and shear modulus G, of the materials under tensile/compression and torsion tests. During these tests, total calculated area under the linear curves was defined as the total potential energy stored in the material. Proportionally, stress development and strains occurred in the structure according to the applied load. Principally, application of the stress distributions should be very slowly; on the other hand, at each incremental loading step, the equilibrium state and its equilibrium equations of the specimen should be satisfied. This controlled operation and action-reaction principle have worked under the control mechanism of the testing machine. The total work done by incremental external forces "dW" was equal to total potential energy stored incrementally "dU" in the structure of linear elastic region. Using this principle, the governing equations were satisfied by $dW - dU = 0$. Otherwise, in the case which used high strain rates $\dot{\varepsilon}$, the material behavior would have been examined in the material nonlinearity concept. In the nonlinear elastic material experimental tests, the resulting stress-strain curves represented the combination of the behavior of nonlinear continuous or multiple nonlinear continuous forms. In nonlinear curves, the stored potential energy "U" developed in the elastic limit range was calculated in the consideration of two areas: the first area under the $\sigma - \varepsilon$ nonlinear curve described as the stored potential energy by strain increments $\varepsilon + d\varepsilon$ and the second area above the curve, known as the complementary potential energy by stress increments $\sigma + d\sigma$ stored in the material. Both linear and nonlinear elasticity equations were derived according to the assumption that during loading and unloading stages of the experiments, the material stores its potential energy within the molecules and there was no loss of energy. As known in the molecular concept, the binding energy keeps the molecules together at any instant of time, and in the lack of energy loss such as heat or light, there will be no loss in the total mass of the molecular system. This phenomenon shows us that the system, which has no energy loss, does not combine (no binding status) with another solid object or with atoms that oscillates at short distances. Otherwise, in the case of the material decreases in amount as losing its mass as energy in the form of heat or light during the binding process, the removed energy corresponding to the removed mass can be explained by Einstein's equation $E = mc^2$. Here, E is the binding energy, m is the mass change in the system, and c is the speed of light, respectively. The elasticity solutions were grouped in terms of a variety of the material, geometry, and loading types. Generally, the used geometries were selected as bar-, beam-, plate-, and shell-type isotropic or composite-type structures. In order to obtain analytical and numerical solutions, the three-dimensional elasticity problems can be reduced into two-dimensional problems in the consideration of the plane stress and plane strain concepts of the elasticity. By these methods the total number of unknowns will be equal to total numbers of equations. Otherwise, some unknown values will stay in unsolvable or undefined forms. Geometrical, material, and loading symmetries reduce problem-solving difficulties in the analytical and numerical models. On the other hand, continuity conditions in geometries automatically satisfies the continuity conditions in the analytical and numerical solutions of elasticity. For example, the existence of the fourth-order partial derivatives of the assumed solution approximation functions is checking the continuity and compatibility equations. Singularity problems may be discarded by omitting the very small holes, empty spaces, gaps in macroscale, or dislocations and beside these the distances between small particles in microscale. In the case of a three-dimensional problem in elasticity, 15 unknowns were defined as mentioned below. These were six stress components, six strain components, and three displacement components. These unknown values were to be calculated by using 15 elasticity equations, three equilibrium equations, six stress-strain relationships, and six strain-displacement relationships. Continuity conditions were satisfied by considering the six compatibility equations which were derived from 15 elasticity equations in three-dimensional problems. Boundary conditions and the initial conditions were both defined on the boundaries and at the starting time domains, respectively, in order to obtain the solutions under the limitation of approximate and true percentage minimum error calculations. In the case of three-dimensional elasticity problem, 15 unknown values have to be solved by 15 governing equations (the list of the unknowns were six stress components $[\sigma_x \ \sigma_y \ \sigma_z \ \tau_{xy} \ \tau_{yz} \ \tau_{xz}]$ and six strain components $[\varepsilon_x \ \varepsilon_y \ \varepsilon_z \ \gamma_{xy} \ \gamma_{yz} \ \gamma_{xz}]$, and additionally the three displacement components [u v w]) [3, 4]. In solid mechanics and elasticity theory, the governing partial differential equations, the constitutive and kinematics equations, and the initial and boundary conditions have been all defined. However, if at least one of the above conditions has remained partially or entirely unknown, then one has a so-called inverse problem (**Figure 2**) [5]. On the other hand, the elasticity "inverse problem" has been defined for the problems in which they consist of recovering the missing displacements to the solution space corresponding to the applied force data by using the iterative calculation steps. Obviously, lost or uncalculated data developing on one part of a whole domain boundary have directly affected the final configuration of the stress-strain and displacement components and their resultant solution spaces at the other part of this boundary. The proposed solutions were both numerical and analytical (**Figure 2**). Inverse problem of elasticity in other words Cauchy problem (Cauchy-Navier equations of elasticity) has been defined on the accessible outer boundary of the structure. The Cauchy stress tensor components were related with the infinitesimal (incremental calculations) strain tensor components which have been identified in deformed configuration with successive iterations.

The stress-strain relationship in terms of indicial notation is given below:

$$\sigma_{ij} = 2\,\mu\varepsilon_{ij} + \lambda\delta_{ij}\varepsilon_{kk} \tag{1}$$

Here, μ, λ are the Lamé constants. The Cauchy strain components represent the geometrical nonlinearity of the material according to the deformed configuration.

The inverse problem solution depends on the stepwise calculated and so updated Cauchy stress and strain distributions, over the whole boundary of the geometry. Experimentally, tractions and displacements have been measured by nondestructive tests. In isotropic, fiber, and particulate composite material concepts, the stress-strain distributions $\sigma - \varepsilon$ have been examined according to the defined total number of elastic constants in stiffness [C] matrix. The inverse of the stiffness matrix named as the compliance matrix $[S] = [C]^{-1}$ includes the elastic constants in $\varepsilon - \sigma$ strain versus stress equations. In the generalized Hook's law, anisotropic crystalline materials have been defined with 36 constants. Strain energy function has to be used to show that the number of independent material constants can be reduced from 36 to 21. The solution techniques as iterative methods, inverse method, semi-inverse method, variational formulation, finite element method, finite volume method, and meshless method have been listed in **Figure 2**. The experimental solution techniques have been explained by tensile, compression, torsion, impact, and bending mechanical tests. Nondestructive tests (NDT) have been used to obtain informational data from the surfaces of the materials (nanoindentation-hardness testing).

5. Conclusion

In this introduction chapter, the historical development of the elasticity concept and its engineering properties were presented briefly. According to Newton's action and reaction principle, the materials behave linear or nonlinear elastically under typical loading. Elasticity theory provides necessarily required equations and solution techniques. The action-response principle defined between the work done by the forces and the potential energy stored has been explained by the material elastic constants. The mechanical response of a homogeneous isotropic linearly elastic material can be explained by two physical constants, Young's modulus and Poisson's ratio. The elastic properties of particle composites, consisting in a dispersion of nonlinear (spherical or cylindrical) nonhomogeneities into a linear solid matrix, were explained by homogenization procedure. The linear elastic constants of fiber composite materials have been defined according to their three principle directions [6]. These principle directions coincided with the fiber orientations located in each layer. By contrast, the physical-mechanical properties of nonlinear elastic materials have generally been described by parameters which have formations as the scalar functions of the deformation, and their material properties have been determined by selecting the suitable solution techniques.

Author details

Ezgi Günay
Mechanical Engineering Department, Engineering Faculty, Gazi University, Ankara, Turkey

*Address all correspondence to: ezgigunay@gazi.edu.tr

References

[1] Todhunter I. A history of the theory of elasticity and of strength of the materials from Galilei to the present time, Vol. II. Saint-Venant to Lord Kelvin. King's College, Cambridge: Pearson; 1893. 574 p. http://www.archive.org/details/historyoftheoryo02todhuoft

[2] Todhunter I. A history of the theory of elasticity and of strength of the materials from Galilei to the present time. In: Saint-Venant to Lord Kelvin, Part 1. Vol. 2. King's College Cambridge Pearson; 2014. 348 p. DOI: 10.1017/CBO9781107280076

[3] Timoshenko SP, Goodier JN. Theory of Elasticity. 3rd ed. New York: McGraw-Hill; 1970. 580 p

[4] Ugural AC, Fenster S. Advanced Mechanics of Materials and Applied Elasticity. 5th ed. New Jersey: Prentice Hall; 2012. 680 p

[5] Karageorghis A, Lesnic D, Marin L. The method of fundamental solutions for three-dimensional inverse geometric elasticity problems. Computers and Structures. 2016;**166**:51-59. DOI: 10.1016/j.compstruc.2016.01.010

[6] Jones RM. Mechanics of Composite Materials. New York: Hemisphere Publishing Co; 1975. 355 p

Chapter 2

An Overview of Stress-Strain Analysis for Elasticity Equations

Pulkit Kumar, Moumita Mahanty and Amares Chattopadhyay

Abstract

The present chapter contains the analysis of stress, analysis of strain and stress-strain relationship through particular sections. The theory of elasticity contains equilibrium equations relating to stresses, kinematic equations relating to the strains and displacements and the constitutive equations relating to the stresses and strains. Concept of normal and shear stresses, principal stress, plane stress, Mohr's circle, stress invariants and stress equilibrium relations are discussed in analysis of stress section while strain-displacement relationship for normal and shear strain, compatibility of strains are discussed in analysis of strain section through geometrical representations. Linear elasticity, generalized Hooke's law and stress-strain relations for triclinic, monoclinic, orthotropic, transversely isotropic, fiber-reinforced and isotropic materials with some important relations for elasticity are discussed.

Keywords: analysis of stress, analysis of strain, Mohr's circle, compatibility of strain, stress-strain relation, generalized Hooke's law

1. Introduction

If the external forces producing deformation do not exceed a certain limit, the deformation disappears with the removal of the forces. Thus the elastic behavior implies the absence of any permanent deformation. Every engineering material/ composite possesses a certain extent of elasticity. The common materials of construction would remain elastic only for very small strains before exhibiting either plastic straining or brittle failure. However, natural polymeric composites show elasticity over a wider range and the widespread use of natural rubber and similar composites motivated the development of finite elasticity. The mathematical theory of elasticity is possessed with an endeavor to decrease the computation for condition of strain, or relative displacement inside a solid body which is liable to the activity of an equilibrating arrangement of forces, or is in a condition of little inward relative motion and with tries to obtain results which might have been basically essential applications to design, building, and all other helpful expressions in which the material of development is solid.

The elastic properties of continuous materials are determined by the underlying molecular structure, but the relation between material properties and the molecular structure and arrangement in materials is complicated. There are wide classes of materials that might be portrayed by a couple of material constants which can be determined by macroscopic experiments. The quantity of such constants relies upon the nature of the crystalline structure of the material. In this section, we give a short but then entire composition of the basic highlights of applied elasticity having pertinence to our topics. This praiseworthy theory, likely the most successful and best surely understood theory of elasticity, has been given numerous excellent and comprehensive compositions. Among the textbooks including an ample coverage of the problems, we deal with in this chapter which are discussed earlier by Love [1], Sokolnikoff [2], Malvern [3], Gladwell [4], Gurtin [5], Brillouin [6], Pujol [7], Ewing, Jardetsky and Press [8], Achenbach [9], Eringen and Suhubi [10], Jeffreys and Jeffreys [11], Capriz and Podio-Guidugli [12], Truesdell and Noll [13] whose use of direct notation and we find appropriate to avoid encumbering conceptual developments with component-wise expressions. Meriam and Kraige [14] gave an overview of engineering mechanics in theirs book and Podio-Guidugli [15, 16] discussed the strain and examples of concentrated contact interactions in simple bodies in the primer of elasticity. Interestingly, no matter how early in the history of elasticity the consequences of concentrated loads were studied, some of those went overlooked until recently [17–22]. The problem of the determination of stress and strain fields in the elastic solids are discussed by many researchers [23–33]. Belfield et al. [34] discussed the stresses in elastic plates reinforced by fibers lying in concentric circles. Biot [35–38] gave the theory for the propagation of elastic waves in an initially stressed and fluid saturated transversely isotropic media. Borcherdt and Brekhovskikh [39–41] studied the propagation of surface waves in viscoelastic layered media. The fundamental study of seismic surface waves due to the theory of linear viscoelasticity and stress-strain relationship is elaborated by some notable researchers [42–46]. The stress intensity factor is computed due to diffraction of plane dilatational waves by a finite crack by Chang [47], magnetoelastic shear waves in an infinite self-reinforced plate by Chattopadhyay and Choudhury [48]. The propagation of edge wave under initial stress is discussed by Das and Dey [49] and existence and uniqueness of edge waves in a generally anisotropic laminated elastic plates by Fu and Brookes [50, 51]. The basic and historical literature about the stress-strain relationship for propagation of elastic waves in kinds of medium is given by some eminent researchers [52–57]. Kaplunov, Pichugin and Rogersion [58–60] have discussed the propagation of extensional edge waves in in semi-infinite isotropic plates, shells and incompressible plates under the influence of initial stresses. The theory of boundary layers in highly anisotropic and/or reinforced elasticity is studied by Hool, Kinne and Spencer [61, 62].

This chapter addresses the analysis of stress, analysis of strain and stress-strain relationship through particular sections. Concept of normal and shear stress, principal stress, plane stress, Mohr's circle, stress invariants and stress equilibrium relations are discussed in analysis of stress section while strain-displacement relationship for normal and shear strain, compatibility of strains are discussed in analysis of strain section through geometrical representations too. Linear elasticity generalized Hooke's law and stress-strain relation for triclinic, monoclinic, orthotropic, transversely isotropic and isotropic materials are discussed and some important relations for elasticity are deliberated.

2. Analysis of stress

A body consists of huge number of grains or molecules. The internal forces act within a body, representing the interaction between the grains or molecules of the body. In general, if a body is in statically equilibrium, then the internal forces are

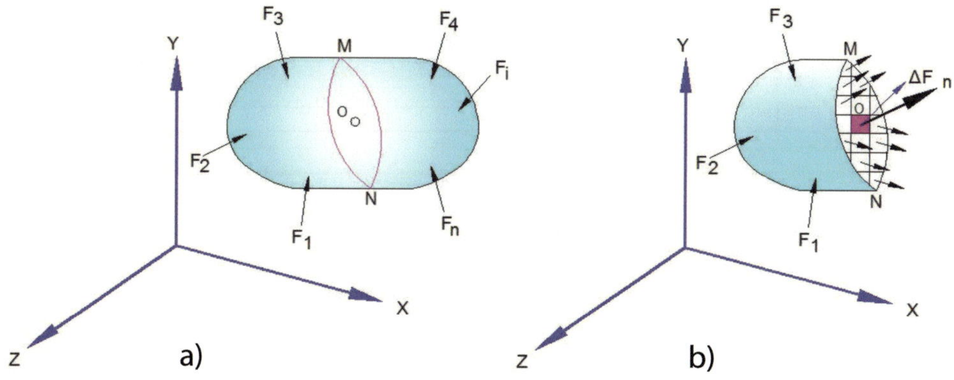

Figure 1.
Forces acting on a (a) body, (b) cross-section of the body.

equilibrated on the basis of Newton's third law. The internal forces are always present even though the external forces are not active.

To examine these internal forces at a point O in **Figure 1(a)**, inside the body, consider a plane MN passing through the point O. If the plane is divided into a number of small areas, as in the **Figure 1(b)**, and the forces acting on each of these are measured, it will be observed that these forces vary from one small area to the next. On the small area ΔA at point O, a force ΔF will be acting as shown in **Figure 1(b)**. From this the concept of stress as the internal force per unit area can be understood. Assuming that the material is continuous, the term "stress" at any point across a small area ΔA can be defined by the limiting equation as below.

$$\textbf{Stress}\ (\sigma) = \lim_{\Delta A \to 0} \frac{\Delta F}{\Delta A} \tag{1}$$

where ΔF is the internal force on the area ΔA surrounding the given point. Forces which act on an element of material may be of two types:

i. body forces and

ii. surface forces.

Body forces always act on every molecule of a body and are proportional to the volume whereas surface force acts over the surface of the body and is measure in terms of force per unit area. The force acting on a surface may resolve into normal stress and shear stress. Normal stress may be tensile or compressive in nature. Positive side of normal stress is for tensile stress whilst negative side is for compressive.

2.1 Concept of normal stress and shear stress

Figure 2(a) shows the rectangular components of the force vector ΔF referred to corresponding axes. Taking the ratios $\Delta F_x/\Delta A_x$, $\Delta F_y/\Delta A_x$, $\Delta F_z/\Delta A_x$, three quantities that set up the average intensity of the force on the area ΔA_x When the limit $\Delta A \to 0$, the above ratios are characterized as the force intensity acting on X-face at point O. These values associated with three intensities are defined as the "Stress components" related with the X-face at point O. The stress component parallel to the surface are called "Shear stress component," is indicated by τ. The

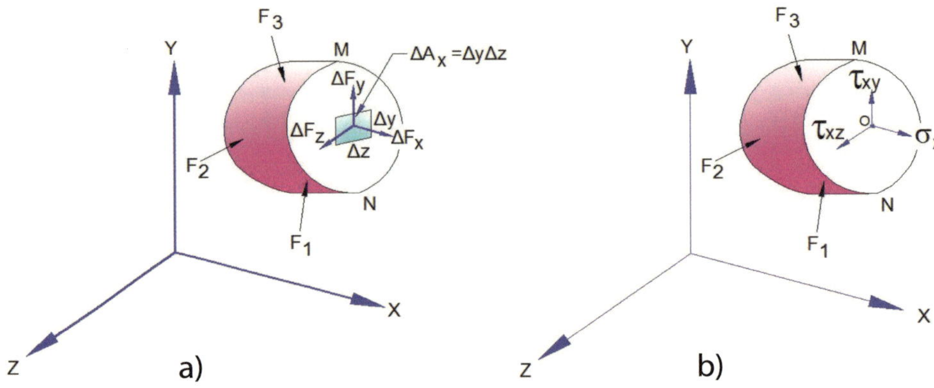

Figure 2.
(a) Force components of ΔF acting on small area centered at point O and (b) stress components at point O.

shear stress component acting on the X-face in the Y-direction is identified as τ_{xy}. The stress component perpendicular to the face is called "Normal Stress" or "Direct stress" component and is denoted by σ.

From the above discussions, the stress components on the X-face at point O are defined as follows in terms of force intensity ratios

$$\left.\begin{aligned} \sigma_x &= \lim_{\Delta A_x \to 0} \frac{\Delta F_x}{\Delta A_x} \\ \tau_{xy} &= \lim_{\Delta A_x \to 0} \frac{\Delta F_y}{\Delta A_x} \\ \tau_{xz} &= \lim_{\Delta A_x \to 0} \frac{\Delta F_z}{\Delta A_x} \end{aligned}\right\} \tag{2}$$

and the above stress components are illustrated in **Figure 2(b)**.

2.2 Stress components

Three mutually perpendicular coordinate axes x, y, z are taken. We consider the stresses act on the surface of the cubic element of the substance. When a force is applied, as mean that the state of stress is perfectly homogeneous throughout the element and that the body is in equilibrium as shown in **Figure 3**. There are nine quantities which are acting on the faces of the cubic and are known as the stress components.

In matrix notation, the stress components can be written as

$$\begin{pmatrix} \sigma_x & \tau_{xy} & \tau_{xz} \\ \tau_{yx} & \sigma_y & \tau_{yz} \\ \tau_{zx} & \tau_{zy} & \sigma_z \end{pmatrix} \tag{3}$$

which completely define the state of stress in the elemental cube. The first suffix of the shear stress refers to the normal to the plane on which the stress acts and the second suffix refer to the direction of shear stress on this plane. The nine stress components which are derived in matrix form are not all independent quantities.

2.3 Principal stress and stress invariants

Let us consider three mutually perpendicular planes in which shear stress is zero and on these planes the normal stresses have maximum or minimum values. These

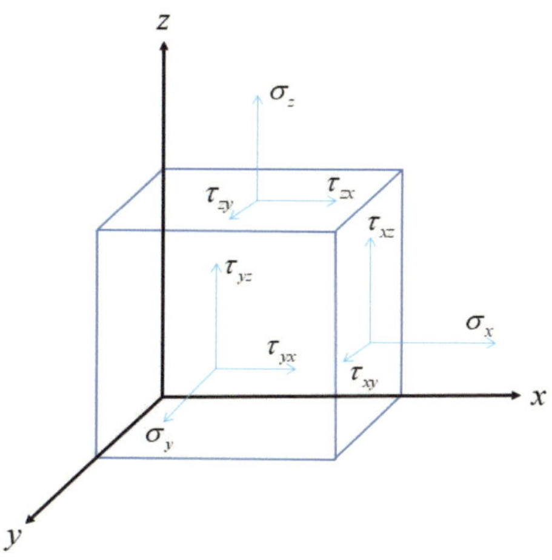

Figure 3.
Stress components acting on cube.

normal stresses are referred to as principal stresses and the plane in which these normal stresses act is called principal plane.

Invariants mean those amounts that are unexchangeable and do not differ under various conditions. With regards to stress components, invariants are such quantities that don't change with rotation of axes or which stay unaffected under transformation, from one set of axes to another. Subsequently, the combination of stresses at a point that don't change with the introduction of co-ordinate axis is called stress invariants.

2.4 Plane stress

Numerous metal shaping procedures include biaxial condition of stress. On the off chance that one of the three normal and shear stresses acting on a body is zero, the state of stress is called plane stress condition. All stresses act parallel to x and y axes. Plane pressure condition is gone over in numerous engineering and forming applications. Regularly, slip can be simple if the shear stress following up on the slip planes is adequately high and acts along favored slip direction. Slip planes may be inclined with respect to the external stress acting on solids. It becomes necessary to transform the stresses acting along the original axes into the inclined planes. Stress change ends up essential in such cases.

2.4.1 Stress transformation in plane stress

Consider the plane stress condition acting on a plane as shown in **Figure 4**. Let us investigate the state of stresses onto a transformed plane which is inclined at an angle θ with respect to x, y axes.

Let by rotating of the x and y axes through the angle θ, a new set of axes X' and Y' will be formed. The stresses acting on the plane along the new axes are obtained when the plane has been rotated about the z axis. In order to obtain these transformed stresses, we take equilibrium of forces on the inclined plane both perpendicular to and parallel to the inclined plane.

Thus, the expression for transformed stress using the direction cosines can be written as

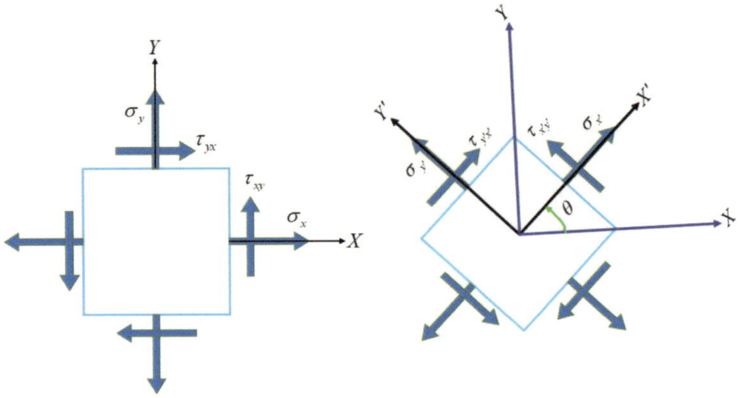

Figure 4.
Representation of stresses on inclined plane.

$$\begin{aligned}\sigma_{x'} &= l_{x'x}^2\sigma_x + l_{x'y}^2\sigma_y + 2l_{x'x}l_{x'y}\tau_{xy} \\ &= 2\cos^2\theta\sigma_x + 2\sin^2\theta\sigma_y + 2\cos\theta\sin\theta\tau_{xy}\end{aligned} \tag{4}$$

Similarly, write for the *y'* normal stress and shear stress.
The transformed stresses are given as

$$\begin{aligned}\sigma_{x'} &= \frac{\sigma_x + \sigma_y}{2} + \frac{\sigma_x - \sigma_y}{2}\cos 2\theta + \tau_{xy}\sin 2\theta \\ \sigma_{y'} &= \frac{\sigma_x + \sigma_y}{2} - \frac{\sigma_x - \sigma_y}{2}\cos 2\theta - \tau_{xy}\sin 2\theta\end{aligned} \tag{5}$$

and

$$\tau_{x'y'} = \frac{\sigma_y - \sigma_x}{2}\sin 2\theta + \tau_{xy}\cos 2\theta$$

where $\sigma_{x'}$ and $\tau_{x'y'}$ are respectively the normal and shear stress acting on the inclined plane. The above three equations are known as transformation equations for plane stress.

In order to design components against failure the maximum and minimum normal and shear stresses acting on the inclined plane must be derived. The maximum normal stress and shear stress can be found when we differentiate the stress transformation equations with respect to θ and equate to zero. The maximum and minimum stresses are known as principal stresses and the plane of acting is named as principal planes.

Maximum normal stress is given by

$$\sigma_1, \sigma_2 = \frac{\sigma_x + \sigma_y}{2} \pm \sqrt{\left(\frac{\sigma_x - \sigma_y}{2}\right)^2 + \tau_{xy}^2} \tag{6}$$

and maximum shear stress is

$$\tau_{\max} = \sqrt{\left(\frac{\sigma_x - \sigma_y}{2}\right)^2 + \tau_{xy}^2} \tag{7}$$

$$\text{with } \tau_{\max} = \frac{\sigma_1 - \sigma_2}{2}. \tag{8}$$

DOI: http://dx.doi.org/10.5772/intechopen.82066

The plane on which the principal normal stress acts, the shear stress is zero and vice versa. The angle corresponding to the principal planes can be obtained from $\tan 2\theta = \frac{\tau_{xy}}{\frac{\sigma_x - \sigma_y}{2}}$ for the principal normal planes and $\tan 2\theta = \frac{\tau_{xy}}{\frac{\sigma_x - \sigma_y}{2}}$ is for the principal shear plane.

2.4.2 Mohr's circle for plane stress

The transformation equations of plane stress which are given by Eq. (5) can be represented in a graphical form (**Figure 5**) by *Mohr's circle*. The transformation equations are sufficient to get the normal and shear stresses on any plane at a point, with Mohr's circle one can easily visualize their variation with respect to plane orientation θ.

2.4.2.1 Equations of Mohr's circle

Rearranging the terms of Eq. (5), we get

$$\sigma_{x'} - \frac{\sigma_x + \sigma_y}{2} = \frac{\sigma_x - \sigma_y}{2} \cos 2\theta + \tau_{xy} \sin 2\theta \tag{9.1}$$

and

$$\tau_{x'y'} = -\left(\frac{\sigma_x - \sigma_y}{2}\right) \sin 2\theta + \tau_{xy} \cos 2\theta \tag{9.2}$$

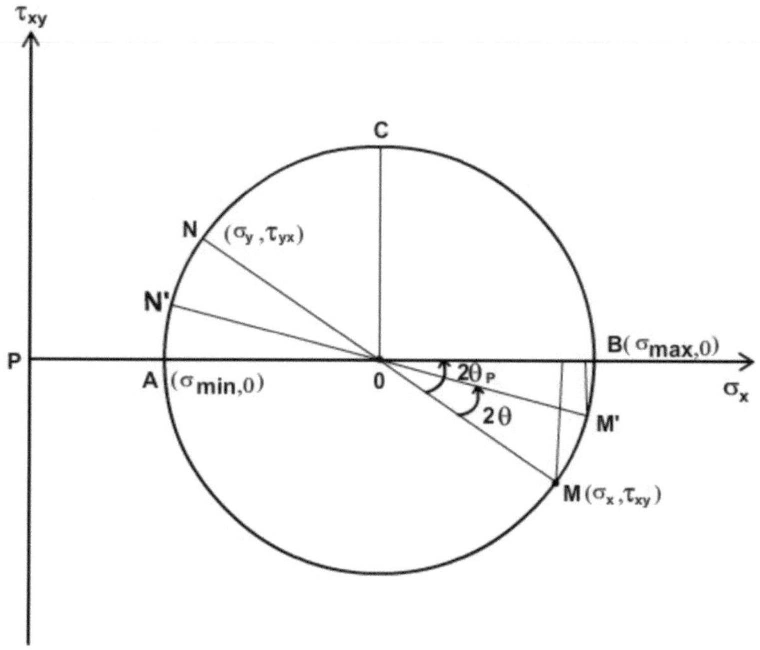

Figure 5.
Mohr's circle diagram.

Squaring and adding the Eqs. (9.1) and (9.2), result in

$$\left(\sigma_{x'} - \frac{\sigma_x + \sigma_y}{2}\right)^2 + \tau_{x'y'}^2 = \left(\frac{\sigma_x - \sigma_y}{2}\right)^2 + \tau_{xy}^2 \tag{10}$$

For simple representation of Eq. (10), the following notations are used

$$\sigma_{av} = \frac{\sigma_x + \sigma_y}{2}, \; r = \left[\left(\frac{\sigma_x - \sigma_y}{2}\right)^2 + \tau_{xy}^2\right]^{1/2} \tag{11}$$

Thus, the simplified form of Eq. (10) can be written as

$$(\sigma_{x'} - \sigma_{av})^2 + \tau_{x'y'}^2 = r^2 \tag{12}$$

Eq. (12) represents the equation of a circle in a standard form. This circle has $\sigma_{x'}$ as its abscissa and $\tau_{x'y'}$ as its ordinate with radius r. The coordinate for the center of the circle is $(\sigma_{av}, 0)$.

Mohr's circle is drawn by considering the stress coordinates σ_x as its abscissa and τ_{xy} as its ordinate, and this plane is known as the stress plane. The plane on the element bounded with xy coordinates in the material is named as physical plane. Stresses on the physical plane M is represented by the point M on the stress plane with σ_x and τ_{xy} coordinates.

Stresses on the physical plane which is normal to i.e. N, is given by the point N on the stress plane with σ_y and τ_{yx}. O is the intersecting point of line MN and which is at the center of the circle and radius of the circle is OM. Now, the stresses on a plane, making θ inclination with x axis in physical plane can be determined as follows.

An important point to be noted here is that a plane which has a θ inclination in physical plane will make 2θ inclination in stress plane M. Hence, rotate the line OM in stress plane by 2θ counter clockwise to obtain the plane M'. The coordinates of M' in stress plane define the stresses acting on plane M' in physical plane and it can be easily verified.

$$\sigma_{x'} = PO + r\cos\left(2\theta_p - 2\theta\right) \tag{13}$$

where $PO = \frac{\sigma_x+\sigma_y}{2}$, $r = \left[\left(\frac{\sigma_x-\sigma_y}{2}\right)^2 + \tau_{xy}^2\right]^{1/2}$, $\cos 2\theta_p = \frac{\sigma_x-\sigma_y}{2r}$, $\sin 2\theta_p = \frac{\tau_{xy}}{2r}$.
On simplifying Eq. (13)

$$\sigma_{x'} = \frac{\sigma_x + \sigma_y}{2} + \frac{\sigma_x - \sigma_y}{2}\cos 2\theta + \tau_{xy}\sin 2\theta \tag{14}$$

Eq. (14) is same as the first equation of Eq. (5).
This way it can be proved for shear stress $\tau_{x'y'}$ on plane M' (do yourself).

2.4.3 Stress equilibrium relation

Let σ_x, τ_{yx}, τ_{zx} are the stress components acting along the x-direction, τ_{xy}, σ_y, τ_{zy} are the stress components acting along the y-direction and τ_{xz}, τ_{yz}, σ_z are the stress components acting along the z-direction. The body forces F_x, F_y, F_z acting along x,

y, z direction respectively. Then the stress equilibrium relation or equation of motion in terms of stress components are given by

$$\left.\begin{aligned} \frac{\partial \sigma_x}{\partial x}+\frac{\partial \tau_{yx}}{\partial y}+\frac{\partial \tau_{zx}}{\partial z}+F_x=0, \\ \frac{\partial \tau_{xy}}{\partial x}+\frac{\partial \sigma_y}{\partial y}+\frac{\partial \tau_{zy}}{\partial z}+F_y=0, \\ \frac{\partial \tau_{xz}}{\partial x}+\frac{\partial \tau_{yz}}{\partial y}+\frac{\partial \sigma_z}{\partial z}+F_z=0. \end{aligned}\right\} \tag{15}$$

3. Analysis of strain

While defining a stress it was pointed out that stress is an abstract quantity which cannot be seen and is generally measured indirectly. Strain differs in this respect from stress. It is a complete quantity that can be seen and generally measured directly as a relative change of length or shape. In generally, stress is the ratio of change in original dimension and the original dimension. It is the dimensionless constant quantity.

3.1 Types of strain

Strain may be classified into three types; normal strain, shear strain and volumetric strain.

The normal strain is the relative change in length whether shearing strain relative change in shape. The volumetric strain is defined by the relative change in volume.

3.2 Strain-displacement relationship

3.2.1 Normal strain

Consider a line element of length Δx emanating from position (x, y) and lying in the x-direction, denoted by AB in **Figure 6**. After deformation the line element occupies $A'B'$, having undergone a translation, extension and rotation.

The particle that was originally at x has undergone a displacement $u_x(x, y)$ and the other end of the line element has undergone a displacement $u_x(x + \Delta x, y)$. By the definition of normal strain

$$\varepsilon_{xx}=\frac{A'B^{*}-AB}{AB}=\frac{u_x(x+\Delta x,y)-u_x(x,y)}{\Delta x}. \tag{16}$$

In the limit $\Delta x \to 0$, Eq. (16) becomes

$$\varepsilon_{xx}=\frac{\partial u_x}{\partial x} \tag{17}$$

This partial derivative is a **displacement gradient**, a measure of how rapid the displacement changes through the material, and is the strain at (x, y). Physically, it represents the (approximate) unit change in length of a line element.

Similarly, by considering a line element initially lying in the y-direction, the strain in the y-direction can be expressed as

$$\varepsilon_{yy}=\frac{\partial u_y}{\partial y}. \tag{18}$$

3.2.2 Shear strain

The particles A and B in **Figure 6** also undergo displacements in the y-direction and this is shown in **Figure 7(a)**. In this case, we have

$$B^*B' = \frac{\partial u_y}{\partial x}\Delta x. \tag{19}$$

A similar relation can be derived by considering a line element initially lying in the y-direction. From the **Figure 7(b)**, we have

$$\theta \approx \tan\theta = \frac{\partial u_y/\partial x}{1 + \partial u_x/\partial x} \approx \frac{\partial u_y}{\partial x} \tag{20}$$

provided that (i) θ is small and (ii) the displacement gradient $\partial u_x/\partial x$ is small. A similar expression for the angle λ can be derived as

$$\lambda \approx \frac{\partial u_x}{\partial y} \tag{21}$$

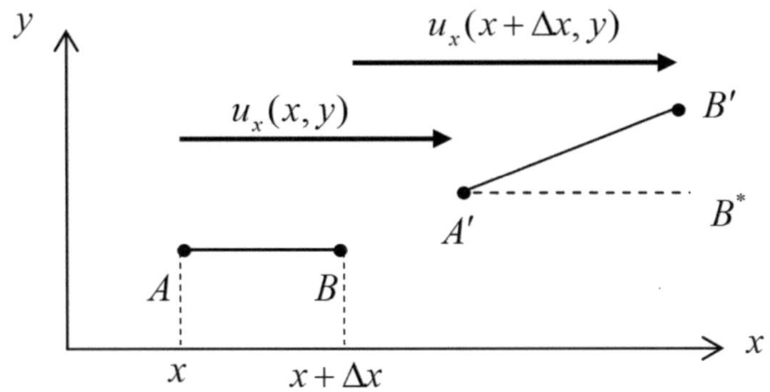

Figure 6.
Deformation of a line element.

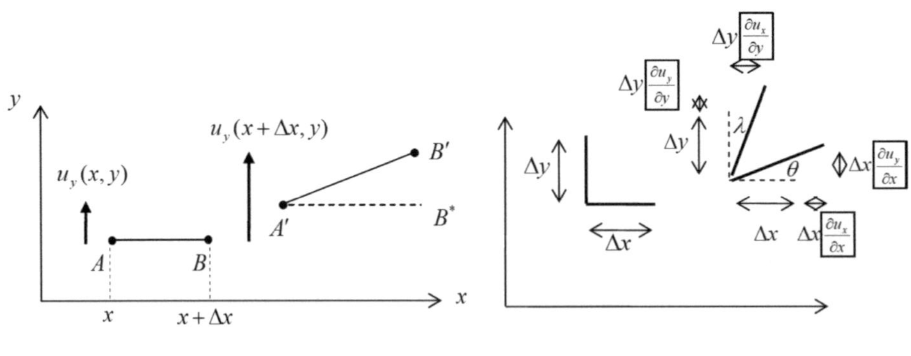

Figure 7.
(a) Deformation of a line element and (b) strains in terms of displacement gradients.

and hence the shear strain can be written in terms of displacement gradients as

$$\varepsilon_{xy} = \frac{1}{2}\left(\frac{\partial u_x}{\partial y} + \frac{\partial u_y}{\partial x}\right). \tag{22}$$

In similar manner, the strain-displacement relation for three dimensional body is given by

$$\begin{aligned} &\varepsilon_{xx} = \frac{\partial u_x}{\partial x}, \ \varepsilon_{yy} = \frac{\partial u_y}{\partial y}, \ \varepsilon_{zz} = \frac{\partial u_z}{\partial z}, \\ &\varepsilon_{xy} = \frac{1}{2}\left(\frac{\partial u_x}{\partial y} + \frac{\partial u_y}{\partial x}\right), \ \varepsilon_{xz} = \frac{1}{2}\left(\frac{\partial u_x}{\partial z} + \frac{\partial u_z}{\partial x}\right), \ \varepsilon_{yz} = \frac{1}{2}\left(\frac{\partial u_y}{\partial z} + \frac{\partial u_z}{\partial y}\right). \end{aligned} \tag{23}$$

3.3 Compatibility of strain

As seen in the previous section, there are three strain-displacement relations Eqs. (17), (18) and (22) but only two displacement components. This implies that the strains are not independent but are related in some way. The relations between the strains are called compatibility conditions.

3.3.1 Compatibility relations

Let us suppose that the point P which is act (x,y) before straining and it will be at P' after straining on the co-ordinate plane Oxy as depicted in **Figure 8**. Then (u,v) is a displacement corresponding to the point P. The variable u and v are the functions of x and y.

Using the fundamental notation

$$\varepsilon_{xx} = \frac{\partial u_x}{\partial x}, \ \varepsilon_{yy} = \frac{\partial u_y}{\partial y}, \ \varepsilon_{xy} = \frac{1}{2}\left(\frac{\partial u_x}{\partial y} + \frac{\partial u_y}{\partial x}\right) \tag{24}$$

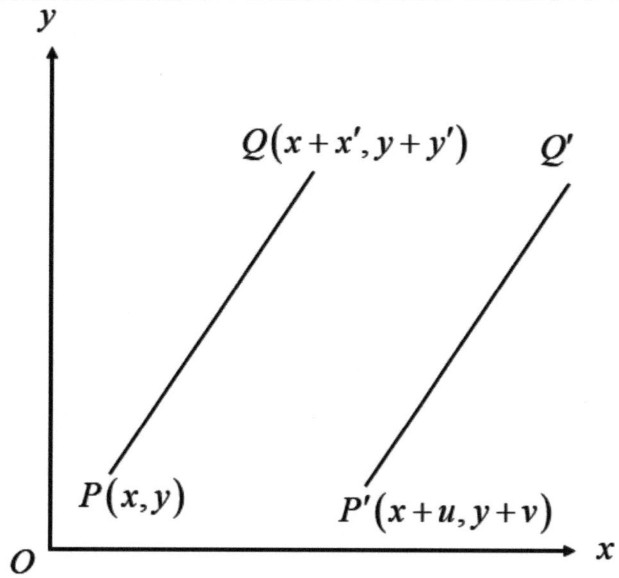

Figure 8.
Deformation of line element.

we get

$$\frac{\partial^2 \varepsilon_{xx}}{\partial y^2} = \frac{\partial^3 u_x}{\partial x \partial y^2}, \frac{\partial^2 \varepsilon_{yy}}{\partial x^2} = \frac{\partial^3 u_y}{\partial x^2 \partial y} \tag{25}$$

$$\frac{\partial^2 \varepsilon_{xy}}{\partial x \partial y} = \frac{1}{2}\left(\frac{\partial^3 u_y}{\partial x^2 \partial y} + \frac{\partial^3 u_x}{\partial x \partial y^2}\right). \tag{26}$$

Eqs. (25) and (26) result in

$$\frac{\partial^2 \varepsilon_{xy}}{\partial x \partial y} = \frac{1}{2}\left(\frac{\partial^2 \varepsilon_{xx}}{\partial y^2} + \frac{\partial^2 \varepsilon_{yy}}{\partial x^2}\right) \tag{27}$$

which is the compatibility condition in two dimension.

4. Stress-strain relation

In the previous section, the state of stress at a point was characterized by six components of stress, and the internal stresses and the applied forces are accompanied with the three equilibrium equation. These equations are applicable to all types of materials as the relationships are independent of the deformations (strains) and the material behavior.

Also, the state of strain at a point was defined in terms of six components of strain. The strains and the displacements are related uniquely by the derivation of six strain-displacement relations and compatibility equations. These equations are also applicable to all materials as they are independent of the stresses and the material behavior and hence.

Irrespective of the independent nature of the equilibrium equations and strain-displacement relations, usually, it is essential to study the general behavior of materials under applied loads including these relations. Strains will be developed in a body due to the application of a load, stresses and deformations and hence it is become necessary to study the behavior of different types of materials. In a general three-dimensional system, there will be 15 unknowns namely 3 displacements, 6 strains and 6 stresses. But we have only 9 equations such as 3 equilibrium equations and 6 strain-displacement equations to achieve these 15 unknowns. It is important to note that the compatibility conditions are not useful for the determination of either the displacements or strains. Hence the additional six equations relating six stresses and six strains will be developed. These equations are known as "Constitutive equations" because they describe the macroscopic behavior of a material based on its internal constitution.

4.1 Linear elasticity generalized Hooke's law

Hooke's law provides the unique relationship between stress and strain, which is independent of time and loading history. The law can be used to predict the deformations used in a given material by a combination of stresses.

The linear relationship between stress and strain is given by

$$\sigma_x = E\varepsilon_{xx} \tag{28}$$

where E is known as Young's modulus.

In general, each strain is dependent on each stress. For example, the strain ε_{xx} written as a function of each stress as

$$\varepsilon_{xx} = C_{11}\sigma_x + C_{12}\sigma_y + C_{13}\sigma_z + C_{14}\tau_{xy} + C_{15}\tau_{yz} + C_{16}\tau_{zx} + C_{17}\tau_{xz} + C_{18}\tau_{zy} + C_{19}\tau_{yx}. \quad (29)$$

Similarly, stresses can be expressed in terms of strains which state that at each point in a material, each stress component is linearly related to all the strain components. This is known as **generalized Hook's law.**

For the most general case of three-dimensional state of stress, Eq. (28) can be written as

$$(\sigma_{ij})_{9\times 1} = (D_{ijkl})_{9\times 9}(\varepsilon_{kl})_{9\times 1} \quad (30)$$

where (D_{ijkl}) is elasticity matrix, (σ_{ij}) is stress components, (ε_{kl}) is strain components.

Since both stress σ_{ij} and strain ε_{ij} are second-order tensors, it follows that D_{ijkl} is a fourth order tensor, which consists of $3^4 = 81$ material constants if symmetry is not assumed.

Now, from $\sigma_{ij} = \sigma_{ji}$ and $\varepsilon_{ij} = \varepsilon_{ji}$, the number of 81 material constants is reduced to 36 under symmetric conditions of $D_{ijkl} = D_{jikl} = D_{ijlk} = D_{jilk}$ which provides stress-strain relation for most general form of anisotropic material.

4.1.1 Stress-strain relation for triclinic material

The stress-strain relation for triclinic material will consist 21 elastic constants which is given by

$$\begin{bmatrix} \sigma_x \\ \sigma_y \\ \sigma_z \\ \tau_{xy} \\ \tau_{yz} \\ \tau_{zx} \end{bmatrix} = \begin{bmatrix} D_{11} & D_{12} & D_{13} & D_{14} & D_{15} & D_{16} \\ D_{12} & D_{22} & D_{23} & D_{24} & D_{25} & D_{26} \\ D_{13} & D_{23} & D_{33} & D_{34} & D_{35} & D_{36} \\ D_{14} & D_{24} & D_{34} & D_{44} & D_{45} & D_{46} \\ D_{15} & D_{25} & D_{35} & D_{45} & D_{55} & D_{56} \\ D_{16} & D_{26} & D_{36} & D_{46} & D_{56} & D_{66} \end{bmatrix} \begin{bmatrix} \varepsilon_{xx} \\ \varepsilon_{yy} \\ \varepsilon_{zz} \\ \varepsilon_{xy} \\ \varepsilon_{yz} \\ \varepsilon_{zx} \end{bmatrix}. \quad (31)$$

4.1.2 Stress-strain relation for monoclinic material

The stress-strain relation for monoclinic material will consist 13 elastic constants which is given by

$$\begin{bmatrix} \sigma_x \\ \sigma_y \\ \sigma_z \\ \tau_{xy} \\ \tau_{yz} \\ \tau_{zx} \end{bmatrix} = \begin{bmatrix} D_{11} & D_{12} & D_{13} & 0 & D_{15} & 0 \\ D_{12} & D_{22} & D_{23} & 0 & D_{25} & 0 \\ D_{13} & D_{23} & D_{33} & 0 & D_{35} & 0 \\ 0 & 0 & 0 & D_{44} & 0 & D_{46} \\ D_{15} & D_{25} & D_{35} & 0 & D_{55} & 0 \\ 0 & 0 & 0 & D_{46} & 0 & D_{66} \end{bmatrix} \begin{bmatrix} \varepsilon_{xx} \\ \varepsilon_{yy} \\ \varepsilon_{zz} \\ \varepsilon_{xy} \\ \varepsilon_{yz} \\ \varepsilon_{zx} \end{bmatrix}. \quad (32)$$

4.1.3 Stress-strain relation for orthotropic material

A material that exhibits symmetry with respect to three mutually orthogonal planes is called an orthotropic material. The stress-strain relation for orthotropic material will consist 9 elastic constants which is given by

$$\begin{bmatrix}\sigma_x\\ \sigma_y\\ \sigma_z\\ \tau_{xy}\\ \tau_{yz}\\ \tau_{zx}\end{bmatrix}=\begin{bmatrix}D_{11} & D_{12} & D_{13} & 0 & 0 & 0\\ D_{12} & D_{22} & D_{23} & 0 & 0 & 0\\ D_{13} & D_{23} & D_{33} & 0 & 0 & 0\\ 0 & 0 & 0 & D_{44} & 0 & 0\\ 0 & 0 & 0 & 0 & D_{55} & 0\\ 0 & 0 & 0 & 0 & 0 & D_{66}\end{bmatrix}\begin{bmatrix}\varepsilon_{xx}\\ \varepsilon_{yy}\\ \varepsilon_{zz}\\ \varepsilon_{xy}\\ \varepsilon_{yz}\\ \varepsilon_{zx}\end{bmatrix}. \tag{33}$$

4.1.4 Stress-strain relation for transversely isotropic material

Transversely isotropic material exhibits a rationally elastic symmetry about one of the coordinate axes x, y and z. In such case, the material constants reduce to 5 as shown below

$$\begin{bmatrix}\sigma_x\\ \sigma_y\\ \sigma_z\\ \tau_{xy}\\ \tau_{yz}\\ \tau_{zx}\end{bmatrix}=\begin{bmatrix}D_{11} & D_{12} & D_{13} & 0 & 0 & 0\\ D_{12} & D_{11} & D_{13} & 0 & 0 & 0\\ D_{13} & D_{13} & D_{33} & 0 & 0 & 0\\ 0 & 0 & 0 & (D_{11}-D_{12})/2 & 0 & 0\\ 0 & 0 & 0 & 0 & D_{44} & 0\\ 0 & 0 & 0 & 0 & 0 & D_{44}\end{bmatrix}\begin{bmatrix}\varepsilon_{xx}\\ \varepsilon_{yy}\\ \varepsilon_{zz}\\ \varepsilon_{xy}\\ \varepsilon_{yz}\\ \varepsilon_{zx}\end{bmatrix}. \tag{34}$$

4.1.5 Stress-strain relation for fiber-reinforced material

The constitutive equation for a fiber-reinforced material whose preferred direction is that of a unit vector $\vec{a}$ is

$$\begin{aligned}\tau_{ij} &= \lambda e_{kk}\delta_{ij} + 2\mu_T e_{ij} + \alpha\left(a_k a_m e_{km}\delta_{ij} + e_{kk}a_i a_j\right) + 2(\mu_L - \mu_T)\left(a_i a_k e_{kj} + a_j a_k e_{ki}\right)\\ &\quad + \beta a_k a_m e_{km} a_i a_j; \quad i, j, k, m = 1, 2, 3\end{aligned} \tag{35}$$

where τ_{ij} are components of stress, e_{ij} are components of infinitesimal strain, and a_i the components of $\vec{a}$, which are referred to rectangular Cartesian co-ordinates x_i. The vector $\vec{a}$ may be a function of position. Indices take the value 1, 2 and 3, and the repeated suffix summation convention is adopted. The coefficients λ, μ_L, μ_T, α and β are all elastic constant with the dimension of stress.

4.1.6 Stress-strain relation for isotropic material

For a material whose elastic properties are not a function of direction at all, only two independent elastic material constants are sufficient to describe its behavior completely. This material is called isotropic linear elastic. The stress-strain relationship for this material is written as

$$\begin{bmatrix} \sigma_x \\ \sigma_y \\ \sigma_z \\ \tau_{xy} \\ \tau_{yz} \\ \tau_{zx} \end{bmatrix} = \begin{bmatrix} D_{11} & D_{12} & D_{12} & 0 & 0 & 0 \\ D_{12} & D_{11} & D_{12} & 0 & 0 & 0 \\ D_{12} & D_{12} & D_{11} & 0 & 0 & 0 \\ 0 & 0 & 0 & (D_{11}-D_{12})/2 & 0 & 0 \\ 0 & 0 & 0 & 0 & (D_{11}-D_{12})/2 & 0 \\ 0 & 0 & 0 & 0 & 0 & (D_{11}-D_{12})/2 \end{bmatrix} \begin{bmatrix} \varepsilon_{xx} \\ \varepsilon_{yy} \\ \varepsilon_{zz} \\ \varepsilon_{xy} \\ \varepsilon_{yz} \\ \varepsilon_{zx} \end{bmatrix} \tag{36}$$

which consists only two independent elastic constants. Replacing D_{12} and $D_{12}\,(D_{11}-D_{12})/2$ by λ and μ which are called Lame's constants and in particular μ is also called shear modulus of elasticity, we get

$$\left.\begin{aligned} \sigma_x &= (2\mu+\lambda)\varepsilon_{xx} + \lambda(\varepsilon_{yy}+\varepsilon_{zz}), \\ \sigma_y &= (2\mu+\lambda)\varepsilon_{yy} + \lambda(\varepsilon_{xx}+\varepsilon_{zz}), \\ \sigma_z &= (2\mu+\lambda)\varepsilon_{zz} + \lambda(\varepsilon_{yy}+\varepsilon_{xx}), \\ \tau_{xy} &= \mu\varepsilon_{xy},\ \tau_{yz} = \mu\varepsilon_{yz},\ \tau_{zx} = \mu\varepsilon_{zx}. \end{aligned}\right\} \tag{37}$$

Also, from the above relation some important terms are induced which are as follow

(1) **Bulk modulus:** Bulk modulus is the relative change in the volume of a body produced by a unit compressive or tensile stress acting uniformly over its surface. Symbolically

$$K = \lambda + \frac{2}{3}\mu. \tag{38}$$

(2) **Young's modulus:** Young's modulus is a measure of the ability of a material to withstand changes in length when under lengthwise tension or compression. Symbolically

$$E = \frac{\mu(3\lambda+2\mu)}{\lambda+\mu}. \tag{39}$$

(3) **Poisson's ratio:** The ratio of transverse strain and longitudinal strain is called Poisson's ratio. Symbolically

$$\nu = \frac{\lambda}{2(\lambda+\mu)}. \tag{40}$$

5. Conclusions

This chapter dealt the analysis of stress, analysis of strain and stress-strain relationship through particular sections. Concept of normal and shear stress, principal stress, plane stress, Mohr's circle, stress invariants and stress equilibrium relations are discussed in analysis of stress section while strain-displacement relationship for normal and shear strain, compatibility of strains are discussed in analysis of strain section through geometrical representations. Linear elasticity, generalized Hooke's law and stress-strain relation for triclinic, monoclinic, orthotropic, transversely-isotropic, fiber-reinforced and isotropic materials with some important relations for elasticity are discussed mathematically.

Acknowledgements

The authors convey their sincere thanks to Indian Institute of Technology (ISM), Dhanbad, India for facilitating us with best research facility and provide a Senior Research Fellowship to Mr. Pulkit Kumar and also thanks to DST Inspire India to provide Senior Research Fellowship to Ms. Moumita Mahanty.

Conflict of interest

There is no conflict of interest to declare.

Author details

Pulkit Kumar*, Moumita Mahanty and Amares Chattopadhyay
Department of Applied Mathematics, Indian Institute of Technology (Indian School of Mines), Dhanbad, Jharkhand, India

*Address all correspondence to: pulkitkumar.maths@gmail.com

References

[1] Love AEH. A Treatise on the Mathematical Theory of Elasticity. Cambridge: Cambridge University Press; 1927

[2] Sokolnikoff IS. Mathematical Theory of Elasticity. New York: McGraw-Hill; 1956

[3] Malvern LE. Introduction to the Mechanics of a Continuous Medium. Englewood Cliffs: Prentice-Hall; 1969

[4] Gladwell GM. Contact Problems in the Classical Theory of Elasticity. Netherland: Springer Science & Business Media; 1980

[5] Gurtin ME. The linear theory of elasticity. In: Linear Theories of Elasticity and Thermoelasticity. Berlin, Heidelberg: Springer; 1973. pp. 1-295

[6] Brillouin L. Tensors in Mechanics and Elasticity. New York: Academic; 1964

[7] Pujol J. Elastic Wave Propagation and Generation in Seismology. Cambridge, New York: Cambridge University Press; 2003

[8] Ewing WM, Jardetsky WS, Press F. Elastic Waves in Layered Media. New York: McGraw-Hill; 1957

[9] Achenbach J. Wave Propagation in Elastic Solids. Amsterdam: North-Holland; 1973

[10] Eringen A, Suhubi E. Elastodynamics. Vol. II. New York: Academic; 1975

[11] Jeffreys H, Jeffreys B. Methods of Mathematical Physics. Cambridge: Cambridge University Press; 1956

[12] Capriz G, Podio-Guidugli P. Whence the boundary conditions in modern continuum physics? In: Capriz G, Podio-Guidugli P, editors. Proceedings of the Symposium. Rome: Accademia dei Lincei; 2004

[13] Truesdell C, Noll W. The non-linear field theories of mechanics. In: The Non-Linear Field Theories of Mechanics. Berlin, Heidelberg: Springer; 2004. pp. 1-579

[14] Meriam JL, Kraige LG. Engineering Mechanics. Vol. 1. New York: Wiley; 2003

[15] Podio-Guidugli P. Strain. In: A Primer in Elasticity. Dordrecht: Springer; 2000

[16] Podio-Guidugli P. Examples of concentrated contact interactions in simple bodies. Journal of Elasticity. 2004;**75**:167-186

[17] Schuricht F. A new mathematical foundation for contact interactions in continuum physics. Archive for Rational Mechanics and Analysis. 2007;**184**: 495-551

[18] Aki K, Richards PG. Quantitative Seismology. Vol. 2. San Francisco: W. H. Freeman & Co; 1980

[19] Mase G. Theory and Problems of Continuum Mechanics, Schaum's outline series. New York: McGraw-Hill; 1970

[20] Morse P, Feshbach H. Methods of Theoretical Physics. Vol. 2. New York: McGraw-Hill; 1953

[21] Mahanty M, Chattopadhyay A, Dhua S, Chatterjee M. Propagation of shear waves in homogeneous and inhomogeneous fibre-reinforced media on a cylindrical Earth model. Applied Mathematical Modelling. 2017;**52**: 493-511

[22] Kumar P, Chattopadhyay A, Singh AK. Shear wave propagation due to a point source. Procedia Engineering. 2017;**173**:1544-1551

[23] Chattopadhyay A, Singh P, Kumar P, Singh AK. Study of Love-type wave propagation in an isotropic tri layers elastic medium overlying a semi-infinite elastic medium structure. Waves in Random and Complex Media. 2017;**28** (4):643-669

[24] Chattopadhyay A, Saha S, Chakraborty M. Reflection and transmission of shear waves in monoclinic media. International Journal for Numerical and Analytical Methods in Geomechanics. 1997;**21**(7):495-504

[25] Chattopadhyay A. Wave reflection and refraction in triclinic crystalline media. Archive of Applied Mechanics. 2004;**73**(8):568-579

[26] Chattopadhyay A, Singh AK. Propagation of magnetoelastic shear waves in an irregular self-reinforced layer. Journal of Engineering Mathematics. 2012;**75**(1):139-155

[27] Udias A, Buforn E. Principles of Seismology. Cambridge, New York: Cambridge University Press; 2017

[28] Novotny O. Seismic Surface Waves. Bahia, Salvador: Instituto de Geociencias; 1999

[29] Brillouin L. Wave Propagation and Group Velocity. Academic Press, New York; 1960

[30] Badriev IB, Banderov VV, Makarov MV, Paimushin VN. Determination of stress-strain state of geometrically nonlinear sandwich plate. Applied Mathematical Sciences. 2015;**9**(77–80): 3887-3895

[31] Jaeger JC, Cook NG, Zimmerman R. Fundamentals of Rock Mechanics. Malden, U.S.A: Blackwell Publishing. 2009

[32] Press F. Seismic wave attenuation in the crust. Journal of Geophysical Research. 1964;**69**(20):4417-4418

[33] Ben-Menahem A, Singh SJ. Seismic Waves and Sources. New York: Springer Verlag; 1981

[34] Belfield AJ, Rogers TG, Spencer AJM. Stress in elastic plates reinforced by fibres lying in concentric circles. Journal of the Mechanics and Physics of Solids. 1983;**31**(1):25-54

[35] Biot MA. The influence of initial stress on elastic waves. Journal of Applied Physics. 1940;**11**(8):522-530

[36] Biot MA. Theory of propagation of elastic waves in a fluid-saturated porous solid. I. Low frequency range. The Journal of the Acoustical Society of America. 1956;**28**(2):168-178

[37] Biot MA. Theory of propagation of elastic waves in a fluid-saturated porous solid. II. Higher frequency range. The Journal of the Acoustical Society of America. 1956;**28**(2):179-191

[38] Biot MA, Drucker DC. Mechanics of incremental deformation. Journal of Applied Mechanics. 1965;**32**(1):957.

[39] Borcherdt RD. Rayleigh-type surface wave on a linear viscoelastic half-space. The Journal of the Acoustical Society of America. 1974;**55**(1):13-15

[40] Borcherdt RD. Viscoelastic Waves in Layered Media. Cambridge, New York: Cambridge University Press; 2009

[41] Brekhovskikh L. Waves in Layered Media. New York: Academic Press; 1980

[42] Bullen KE. Compressibility-pressure hypothesis and the Earth's interior. Geophysical Journal International. 1949; **5**:335-368

[43] Bullen KE, Bullen KE, Bolt BA. An Introduction to the Theory of Seismology. Cambridge, New York: Cambridge University Press; 1985

[44] Chapman C. Fundamentals of Seismic Wave Propagation. Cambridge, UK: Cambridge University Press; 2004

[45] Carcione JM. Wave propagation in anisotropic linear viscoelastic media: Theory and simulated wavefields. Geophysical Journal International. 1990; **101**(3):739-750

[46] Chadwick P. Wave propagation in transversely isotropic elastic media—I. Homogeneous plane waves. Proceedings of the Royal Society of London A. 1989; **422**(1862):23-66

[47] Chang SJ. Diffraction of plane dilatational waves by a finite crack. The Quarterly Journal of Mechanics and Applied Mathematics. 1971;**24**(4): 423-443

[48] Chattopadhyay A, Choudhury S. Magnetoelastic shear waves in an infinite self-reinforced plate. International Journal for Numerical and Analytical Methods in Geomechanics. 1995;**19**(4):289-304

[49] Das SC, Dey S. Edge waves under initial stress. Applied Scientific Research. 1970;**22**(1):382-389

[50] Fu YB. Existence and uniqueness of edge waves in a generally anisotropic elastic plate. Quarterly Journal of Mechanics and Applied Mathematics. 2003;**56**(4):605-616

[51] Fu YB, Brookes DW. Edge waves in asymmetrically laminated plates. Journal of the Mechanics and Physics of Solids. 2006;**54**(1):1-21

[52] Sneddon IN. The distribution of stress in the neighbourhood of a crack in an elastic solid. Proceedings of the Royal Society A. 1946;**187**(1009):229-260

[53] Sengupta PR, Nath S. Surface waves in fibre-reinforced anisotropic elastic media. Sadhana. 2001;**26**(4):363-370

[54] Shearer PM. Introduction to Seismology. 2nd ed. Cambridge: Cambridge University Press; 2009

[55] Sheriff RE, Geldart LP. Exploration Seismology. Cambridge, New York: Cambridge University Press; 1995

[56] Kolsky H. Stress Waves in Solids. New York: Dover Publication; 1963

[57] Gubbins D. Seismology and Plate Tectonics. Cambridge, New York: Cambridge University Press; 1990

[58] Kaplunov J, Prikazchikov DA, Rogerson GA. On three-dimensional edge waves in semi-infinite isotropic plates subject to mixed face boundary conditions. The Journal of the Acoustical Society of America. 2005;**118**(5): 2975-2983

[59] Kaplunov J, Pichugin AV, Zernov V. Extensional edge modes in elastic plates and shells. The Journal of the Acoustical Society of America. 2009;**125**(2): 621-623

[60] Pichugin AV, Rogerson GA. Extensional edge waves in pre-stressed incompressible plates. Mathematics and Mechanics of Solids. 2012;**17**(1):27-42

[61] Hool GA, Kinne WS, Zipprodt RR. Reinforced Concrete and Masonry Structures. New York: McGraw-Hill; 1924

[62] Spencer AJM. Boundary layers in highly anisotropic plane elasticity. International Journal of Solids and Structures. 1974;**10**(10):1103-1123

Section 2

Engineering Applications in Theory of Elasticity

Chapter 3

FEA and Experimentally Determination of Applied Elasticity Problem for Fabricating Aspheric Surfaces

Duc-Nam Nguyen

Abstract

The elastic deformation machining method is suitable for fabricating aspheric surfaces that have excellent physical properties of elastic materials. The machining process is carried out with the deformation model without mold. When vacuum pressure is supplied to the workpiece, the top surfaces of workpiece are deformed into aspheric shape. After machining process, the bottom surface will be formed into the aspheric shape and the top surface returns to its original flat surface form due to internal force and bending moments of the material. However, the accuracy will decrease due to the reduced thickness while the vacuum pressure keeps unchanged during machining process. Therefore, it is necessary to carry out the finite element analysis (FEA) to determine the vacuum pressure with corresponding to the reduced thickness. In addition, the mold with its surface approximates and the desired surface form of the lens is also presented. When uniform vacuum pressure is supplied to the workpiece through small holes of the mold, the workpiece will be deformed into aspheric profile as similar to the mold surfaces. In order to improving the form accuracy, the FEA and the experiment are studied for modifying the mold profile to correspond with bending strength of workpiece material.

Keywords: elastic deformation, aspheric surface, glass lapping, glass molding, vacuum pressure, experimental study, finite element analysis

1. Introduction

Nowadays, for managing laser light in sophisticated and compact laser systems, aspheric lenses are the most powerful lenses. In these systems, it is generally accepted that spherical aberration is the most common performance detractor. From the use of spherical surfaces, it is found that they artificially limit focusing and collimating accuracy. In spite of the fact that spherical geometry is not optimal for refracting light that has been known for centuries, the high cost and difficulty of fabricating nonspherical (aspheric) surfaces has inhibited them from a wider use.

Because aspheric surfaces offer advantages such as high resolution, light weight, and low cost, they are widely used in the opto-electronics industry. As aspheric surfaces are more effective in shaping the light than spherical surfaces, they have

IntechOpen

recently been used in measurement instruments, astronomy, and optical lens [1, 2]. **Figure 1** shows some applications which employ aspheric surfaces.

In most general terms, an optical lens can be determined as a refracting device that reconfigures the light wave front incident upon it. The phase, direction of propagation, intensity, and polarization state are the properties of the incident light beam which are influenced by a lens. Surface form and roughness, diameter, subsurface defects generated during the fabrication process, shape accuracy, physical and mechanical properties of the optical material, and other optical conditions, such as the angle of incidence of light beam, absorption, reflection, and environmental influences, are some of the major characteristics that govern the performance of an optical lens [3].

To overcome the aberration problems of spherical lenses, a number of spherical surfaces with different signs of aberrations have to be utilized to balance and minimize the final aberration to obtain high quality images. In principle, the optical system designer can always use enough spherical lenses to simultaneously correct for all of the common optical aberrations in a lens system if the number of elements used in an optical system is not limited. The number of surfaces required to do this may be so large that the resulting lens assembly is excessively large in size and weight, and expensive to produce. In addition, the transmission of the assembly lens may be unduly reduced due to the residual reflections from each surface, and the bulk absorption in each lens.

The usage of aspheric surfaces, both with and without the incorporation of diffractive elements, allows the design and construction of assembly lens with the same or even better optical performance than an equivalent all-spherical system. However, in most cases, with a significant reduction in the number of elements required, there is a significant improvement in the overall lens assembly size, weight, cost, and optical transmission. In many cases, in an optical system, each aspheric surface can be applied to replace at least two other spherical surfaces. Hence, aspheric lenses are more efficient because additional error-correcting lenses are not required. **Figure 2** is an illustration of spherical and aspheric lens systems.

Cutting techniques such as turning and milling processes are usually utilized for the production of aspheric glass lenses as shown in **Figure 3**.

The machining processes, which usually consist of computer numerically controlled (CNC) generators, are employed to machine an aspheric shape on a lens to generate the desired shape. In glass machining, the roughness on a cutting edge has a larger effect on surface finish than that of metal machining. Glass workpiece can be machined without brittle fractures with an undeformed chip thickness less than 1 μm in milling and turning processes [4–7].

a) b) c)

Figure 1.
Application of aspheric surfaces: (a) measurement instruments, (b) astronomy, and (c) optical lens.

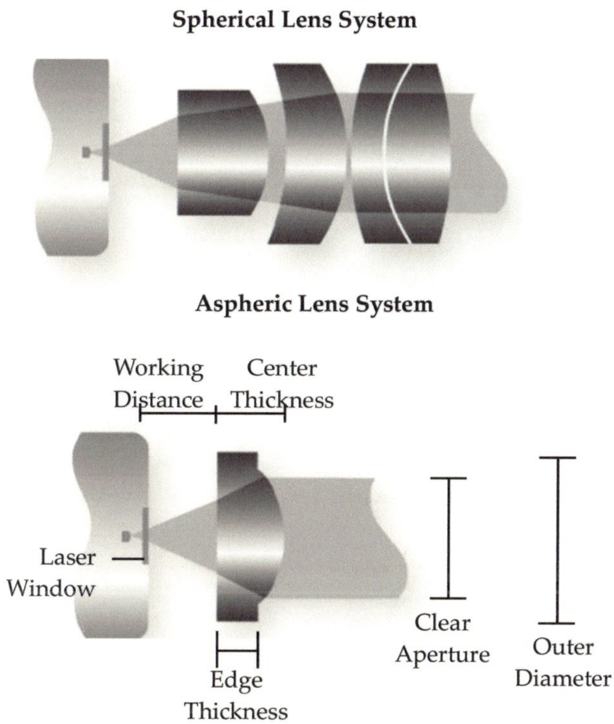

Figure 2.
Spherical vs. aspheric lens systems.

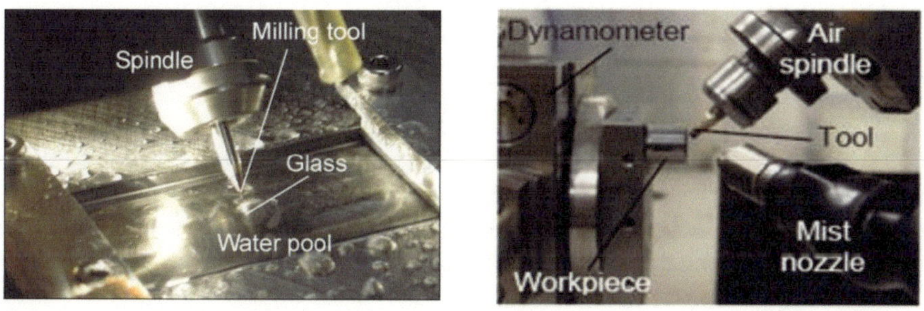

Figure 3.
Schematic illustration of milling and turning processes.

Thereafter, the optical lenses are fine machined by grinding, and then followed by polishing to achieve the good surfaces. In the grinding process, if the depth of cut is below a certain value, the material removal mode is ductile flow which is characterized by low surface roughness and subsurface damage [8–12]. **Figure 4** is an illustration of a precision grinding process.

"Precessions" polishing is an automated polishing method that uses a 7-axis CNC machine tool for polishing spherical and aspheric surfaces [13, 14]. Based on contact between the workpiece surface and polishing tool, the polishing spot of desired size is generated by controlling the load cell in polishing process. The polishing tool then moves in angular steps around the local normal to the part surface during machining process. The 7-axis CNC capability of the machine also makes the generation of free-form surfaces possible. **Figure 5** shows a schematic illustration of a "Precessions" polishing process.

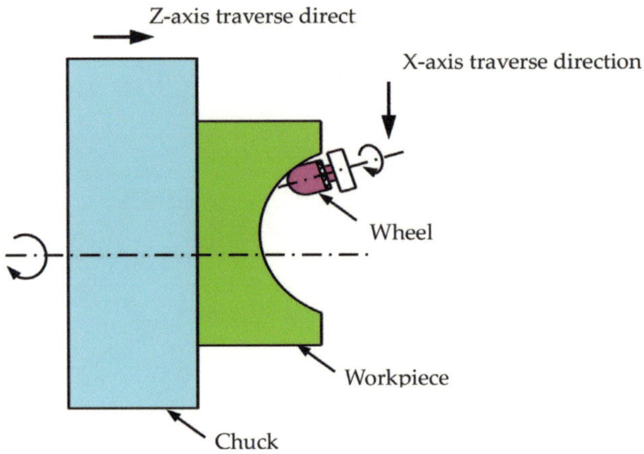

Figure 4.
Schematic illustration of a precision grinding process.

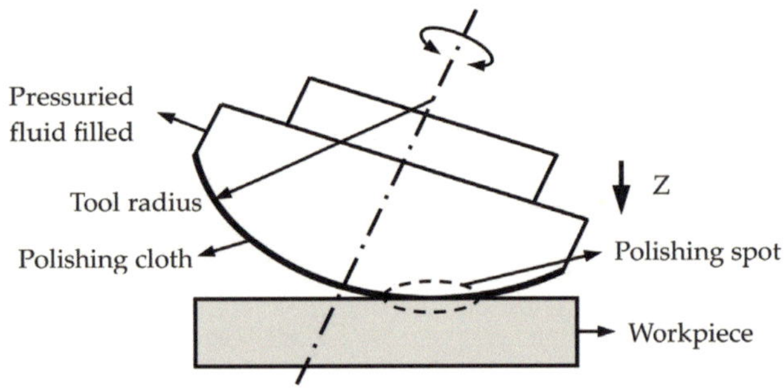

Figure 5.
Schematic illustration of a "Precessions" polishing process.

To fabricate aspheric surfaces, the movement of the tool must be constrained in the machining process. A sub-aperture tool (smaller in size than the lens) on a modified polishing machine is then utilized, and by controlling the amount of time the tool spends working at a given lens location, a desired aspheric surface can be fabricated. In addition to the complexity of the machining processes, conventional aspheric fabrication is highly sensitive to the manufacturing conditions, which strongly depend on the positioning accuracy of the machine, the condition of the grinding wheel, and the vibrations in the system. These factors result in an expensive manufacturing cost and a low production yield.

Compared to traditional cold-working methods, glass molding and precision injection molding have greatly advanced the fabrication technologies for aspheric lens industry because of their unique advantages such as excellent compatibility, high efficiency, great flexibility, and high consistency [15–17]. The mass production of aspheric glass lenses is fabricated by applying the technologies. In the glass molding technique, a glass lens is fabricated by compressing glass melting at a high temperature and replicating the shapes of the mold without any need of further machining. **Figure 6a** shows the process begins by putting a glass gob on top of a lower mold. Both the glass gob and the mold are heated to a molding temperature above the transition temperature of glass (**Figure 6b**). After the glass and the mold temperature have reached a steady state molding temperature, the mold is closed by moving the lower mold (**Figure 6c**). The temperature is maintained during the molding step. All steps are performed in vacuum environment. Then, by holding

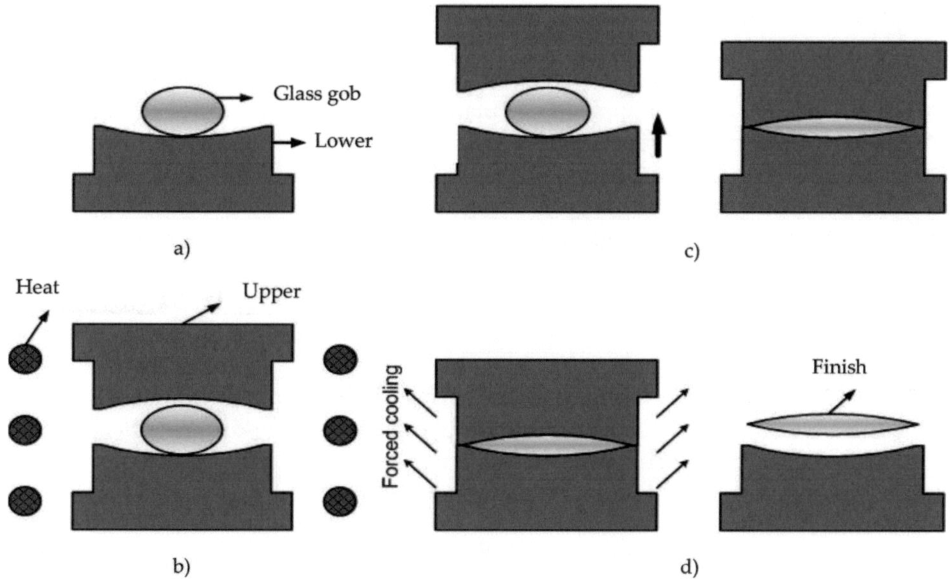

Figure 6.
Schematic illustration of a lens molding process: (a) Molds and glass gob, (b) Heating, (c) Heating and pressing, and (d) Cooling and release.

the pressed load for a short time at a slow cooling rate, the stress in the glass lens is relaxed. Lastly, the formed glass lens is rapidly cooled to ambient temperature and released from the molds (**Figure 6d**). A BK7 glass fabricated using this molding process has surface roughness of approximately 5 nm *Ra*, and form accuracy of 0.2 μm *P-V*.

In spite of the obvious advantages, there are serious drawbacks that currently limit the application of injection molding and glass molding technologies to smaller size aspheric lens fabrications. A typical drawback is the altering of optical properties such as refractive index, due to heating and annealing of the glass material, and the uneven shrinking due to the cooling process that causes error in lens profile [18].

In contrast, the elastic deformation machining method is a good technique that the workpiece will be deformed into aspheric shape prior to the lapping process under vacuum pressure. While the vacuum pressure is remained, the opposite side is polished to optical flatness by the lapping wheel. When the vacuum pressure is released, the bottom surface of the workpiece will be shaped into an aspheric shape and the top surface will restore to its flat surface form by internal force and bending moments. Consequently, for machining materials which have excellent physical properties due to their perfect crystal structures, the elastic deformation method is appropriate [19].

2. Theory of elastic deformation

Based on the elasticity of the material, the circular flat plate is deformed to an aspheric surface by applying the pressure in the elastic deformation machining method. The deflection of the circular plate can be calculated by using appropriate plate theory. There are two types of edge support for circular plate, such as fixed (or clamped) edge and simply supported edge which are considered in this section.

2.1 Basic equations for circular plate in elastic deformation

The amount of deflection of circular plate can be determined by solving the differential equations of an appropriate plate theory [20]. Two types of edge support include clamped and simply supported edge which are used in the elastic deformation method. In the case of simple bending of circular plate, the amount of deflection w is assumed to be very small in comparison with plate thickness. According to the small deflection theory of thin homogenous elastic plates, the deformation in the middle plane of the plate can be neglected and the straight line initially normal to the middle surface to the plate remains straight. In addition, the stress (i.e., transverse normal stress) is small when compared to other stress components and should be neglected in stress-strain relationship. Under these conditions, the three dimensional plate problem can be reduced to two dimensions. The linear theory of elasticity can be used to derive the governing differential equation for a plate subject to uniform transverse loads. The equation for small deformation w of a thin circular plate of constant thickness h is:

$$D.\nabla^2(\nabla^2 \mathrm{w}) - p = 0; D = Eh^3/12(1-\nu^2) \tag{1}$$

E and v are the Young's modulus and Poisson's coefficient. D is the rigidity constant of the plate, and p is the load on the plate. Because of the rotational symmetry, the Laplacian operator ∇^2 in polar coordinates r and θ can be written as:

$$\nabla^4 w = \left(\frac{\partial^2}{\partial r^2} + \frac{1}{r}\frac{\partial}{\partial r} + \frac{1}{r^2}\frac{\partial^2}{\partial \theta^2}\right)\left(\frac{\partial^2 w}{\partial r^2} + \frac{1}{r}\frac{\partial w}{\partial r} + \frac{1}{r^2}\frac{\partial^2 w}{\partial \theta^2}\right) \tag{2}$$

The moment can be written in the form,

$$M_r = -D\left[\frac{\partial^2 w}{\partial r^2} + \nu\left(\frac{1}{r^2}\frac{\partial^2 w}{\partial \theta^2} + \frac{1}{r}\frac{\partial w}{\partial r}\right)\right] \tag{3}$$

$$M_t = -D\left[\frac{1}{r}\frac{\partial w}{\partial r} + \frac{1}{r^2}\frac{\partial^2 w}{\partial \theta^2} + \nu\frac{\partial^2 w}{\partial r^2}\right] \tag{4}$$

where M_r and M_t are radial moment and tangential moment.

If the load acting on the plate is symmetrically distributed about the axis perpendicular to the middle plane of the plate, the deflection w is independent of θ, when Eqs. (3) and (4) becomes:

$$\left(\frac{d}{dr^2} + \frac{1}{r}\frac{d}{dr}\right)\left(\frac{d^2 w}{dr^2} + \frac{1}{r}\frac{dw}{dr}\right) = \frac{p}{D} \tag{5}$$

In other form, it shown as

$$\left(\frac{d^2 w}{dr^2} + \frac{1}{r}\frac{dw}{dr}\right) = \frac{1}{r}\frac{d}{dr}\left(r\frac{dw}{dr}\right) \tag{6}$$

Eq. (5) can be written as

$$\frac{1}{r}\frac{d}{dr}\left\{r\frac{d}{dr}\left[\frac{1}{r}\frac{d}{dr}\left(r\frac{dw}{dr}\right)\right]\right\} = \frac{p}{D} \tag{7}$$

Multiply both sides of Eq. (6) by r and then integrate to obtain,

$$r\frac{d}{dr}\left[\frac{1}{r}\frac{d}{dr}\left(r\frac{dw}{dr}\right)\right]=\frac{pr^2}{2D}+C_1 \tag{8}$$

or,

$$\frac{d}{dr}\left[\frac{1}{r}\frac{d}{dr}\left(r\frac{dw}{dr}\right)\right]=\frac{pr}{2D}+\frac{C_1}{r} \tag{9}$$

By successive integrations, the deflection can arrive finally at

$$w_r=\frac{pr^4}{64D}+C_1r^2\ln r+C_2r^2+C_3\ln r+C_4 \tag{10}$$

The shears for the symmetrically loaded plate can be given as follow,

$$\frac{d}{dr}\left(\frac{d^2w}{dr^2}+\frac{1}{r}\frac{dw}{dr}\right)=-\frac{Q}{D} \tag{11}$$

and from Eq. (9),

$$\frac{pr}{2D}+\frac{C_1}{r}=-\frac{Q}{D} \tag{12}$$

This equation indicates that Q would approach infinity as r approaches zero. To prevent this from happening, we make $C_1 = 0$. From Eq. (10), it can be seen that w becomes infinity at $r = 0$. To avoid this, the constant C_3 must be zero. Thus,

$$w_r=\frac{pr^4}{64D}+C_2r^2+C_4 \tag{13}$$

From Eq. (13), the amount of deflection w_r is the function of r in cylindrical coordinate system.

2.2 Circular plate with simply supported edge

The boundary conditions are $w = 0$ and $M_r = 0$ at $r = a$.
Eq. (13) can be written,

$$\frac{pa^4}{64D}+C_2a^2+C_4=0 \tag{14}$$

and

$$\frac{pa^2}{64}(12+4\nu)+\frac{C_2}{2}(1+\nu)=0 \tag{15}$$

From the equations, we can find

$$C_2=-\frac{pa^2}{32D}\frac{3+\nu}{1+\nu};C_4=\frac{pa^4}{64D}\frac{5+\nu}{1+\nu} \tag{16}$$

so that the deflection of every radial location can be calculated using,

$$w_r = \frac{p(a^2 - r^2)}{64D}\left[\frac{5+\nu}{1+\nu}a^2 - r^2\right] \tag{17}$$

The maximum deflection which occurs at $r = 0$, is given by,

$$w_0 = \frac{pa^4}{64D}\left(\frac{5+\nu}{1+\nu}\right) \tag{18}$$

Substitute $D = Eh^3/12(1 - v^2)$ into Eq. (18), we have:

$$w_0 = \frac{3p(1-\nu)(5+\nu)a}{16E}\left(\frac{a}{h}\right)^3 \tag{19}$$

From Eq. (19), we can see that the maximum deflection of the circular plate is relative of the diameter a and the ratio between diameter a and thickness h.

3. Elastic deformation machining method without mold

Figure 7 shows a schematic illustration of a lens elastic deformation process without mold.

Two surfaces of the workpiece are polished to certain flatness before fabricating as shown in **Figure 7a**. When vacuum pressure is supplied to the workpiece through

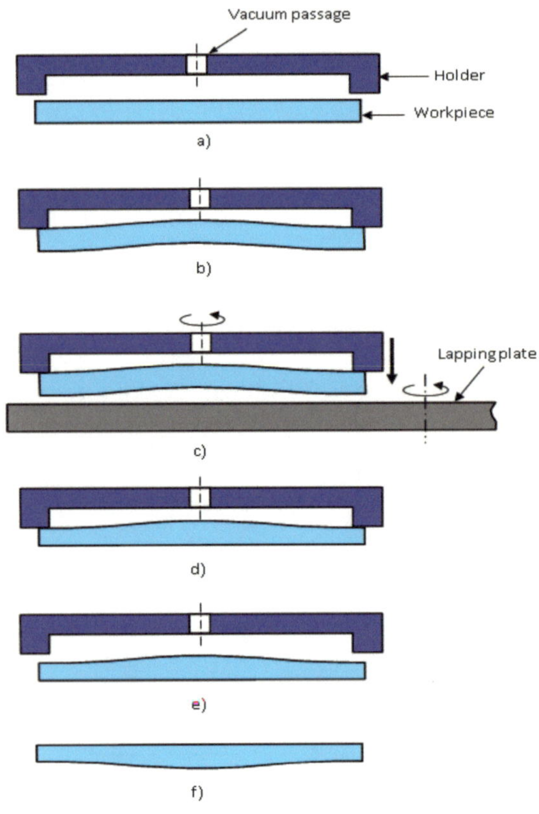

Figure 7.
(a–f) Schematic illustration of an aspheric surface elastic deformation process.

a hole, the workpiece is deformed in the middle. The edge of the workpiece is supported by the holder; therefore, it will be not moved. This makes the workpiece become a formed aspheric shape as presented in **Figure 7b**. The deflection of the workpiece can be calculated by using theoretical equation in boundary conditions of circular plate with simply supported edge. While the vacuum pressure is still remained, the workpiece and the holder start rotating and moving downward in contact with the lapping plate. Its opposite side will be polished to optical flatness as illustrated in **Figure 7c** and **d**. Then, the vacuum pressure is not supplied and the workpiece is also released from the holder as shown in **Figure 7e**. According to the **Figure 7f**, the bottom surface will be formed into the aspheric shape and the top surface returns to its original flat surface form due to material elasticity. It can be seen that the deformed workpiece surfaces can be restored by internal force and bending moments which are created from the vacuum pressure during machining process.

3.1 Finite element analysis for elastic deformation machining process

The vacuum pressure affects an amount of elastic deformation of the workpiece; hence, the accuracy of manufactured profile will also be highly dependent on the vacuum pressure as well. **Figure 8a** and **b** illustrates that the workpiece is lapped and polished to a flat surface while the vacuum pressure stays it at the initial deformed state.

The manufactured workpiece accuracy can be improved by adjusting the vacuum pressure during the machining process because of the changed workpiece thickness [21]. The vacuum pressure is defined by finite element analysis (FEA) results because theoretical calculation for complex surface is more difficult. In simulation process, a circular plate B270 optical glass with the edge supported by the holding device is listed in **Table 1**.

All elements of modeling were created by meshing with A20-node quadratic brick elements in reduced integration (C3D20R). **Figure 9** demonstrates the finite element model as follows.

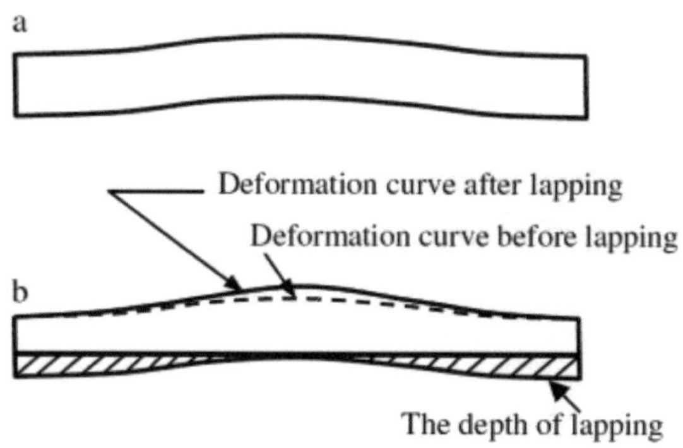

Figure 8.
(a and b) The deforming and lapping processes of glass plate. (a) The glass plate is deformed before lapping and (b) the glass plate is deformed in lapping.

Density (kg/m^3)	Young's modulus (GPa)	Knoop hardness HK100 (kg/mm^2)	Poisson ratio
2550	71.5	542	0.22

Table 1.
Material properties of B270 optical glass [22].

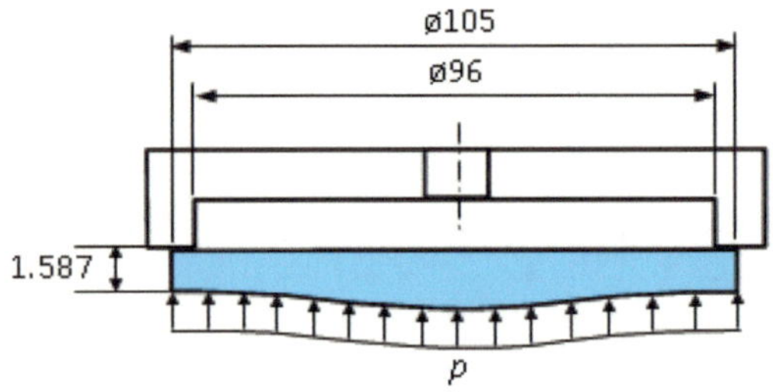

Figure 9.
Simulation model of workpiece.

When the vacuum pressure keeps unchanged, the workpiece thickness is reduced during the lapping process. Therefore, the surface form of a glass plate will have some errors compared to desired surface form at the end of the machining process. The results of the FEA indicate that the deflection of workpiece is greater than desired curve. In order to enhance its accuracy, the vacuum pressure should be fixed at 42 kPa as shown in **Figure 10**.

3.2 Experimental setup

The B270 optical glass which is a clear, high transmission and high purity raw materials is chosen in this experiment. The workpiece sides are lapped and polished to flat surfaces. The lapping process is through the relative motion between the lapping plate and the workpiece, affected by abrasive slurry under distribution load. The silicon carbide (SiC) and cerium oxide (CeO_2) abrasive grain slurry are used in the experiment. The principle of lapping process can be seen in **Figure 11**.

In lapping process, a rigid iron surface covered by a flannelette plate is moved under the load on the glass surface, with abrasive particles suspended in water between them. **Table 2** demonstrates parameters for the machining process. To remove microcrack layer and trace after the lapping process, a polishing step is required. This step is also carried out with the Nanopoli-100 precision polishing machine. The polishing parameters fixed unchanged as that in the initial lapping step, except that the abrasive is changed from SiC to CeO_2 as a fine polishing step.

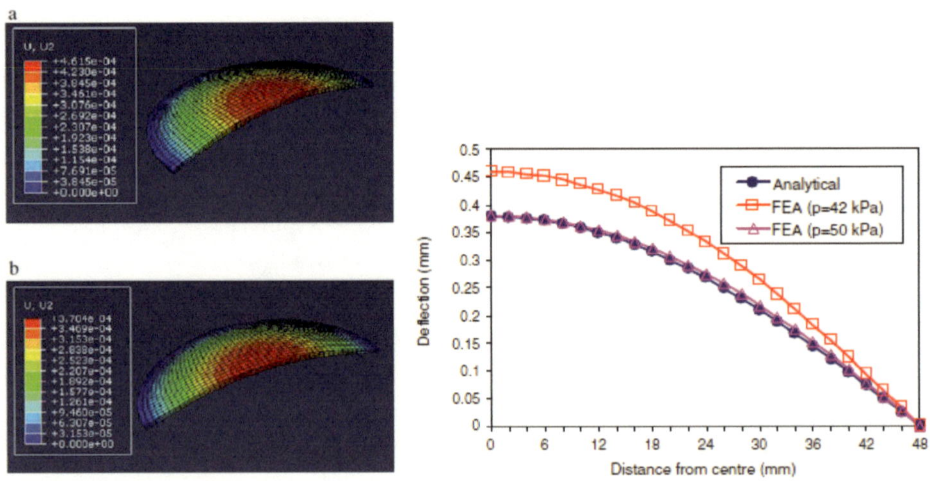

Figure 10.
Finite element analysis results and analytical results: (a) P = 50 kPa and (b) P = 42 kPa.

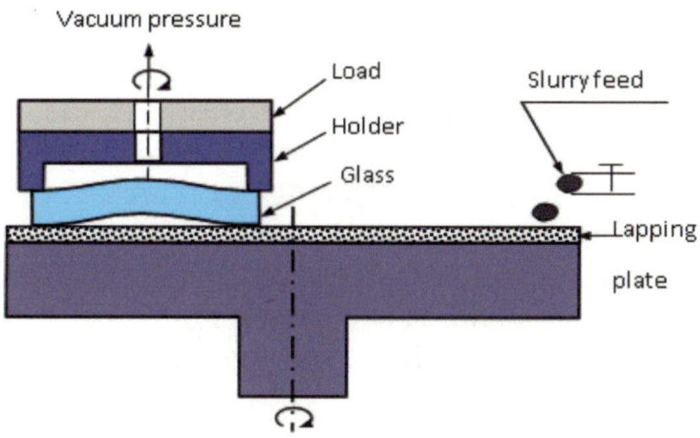

Figure 11.
Principle of lapping and polishing process.

Items	Lapping	Polishing
Abrasive	#1000 SiC	#10,000 CeO_2
Abrasive concentration in slurry (wt%)	10%	10%
Machining load (N)	20	15
Rotating speed of lapping plate (rpm)	60	40
Machining time (min)	240	60

Table 2.
Lapping and polishing conditions.

3.3 Experimental results

The component accuracy can be improved by adjusting the vacuum pressure values to compensate for its lost thickness during the lapping step. The vacuum pressure is defined through FEA results. **Figure 10** shows that the deformation curve of the workpiece is close to the desired curve when the vacuum pressure is fixed at 42 kPa. Therefore, the vacuum pressure should be reduced from 50 to 42 kPa in the experiment. **Figure 12** illustrates the deflection and deviation results of the experimental results and the theoretical calculations.

Depending on reducing pressure from 50 to 42 kPa and keeping stable through the entire lapping step, the experimental results agree greatly with theoretical calculations. The peak-valley value is reached at 1.6 μm.

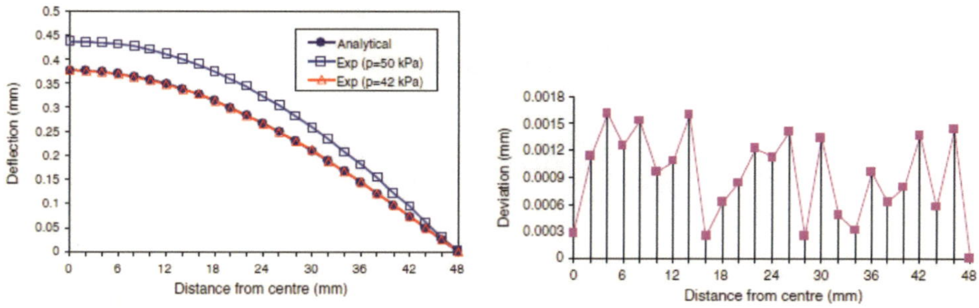

Figure 12.
The experimental results of deflection and deviation against theoretical results.

4. Elastic deformation machining method with mold

In the elastic deformation machining process without mold, the thickness of the plate is reduced while the vacuum pressure remains unchanged. Thus, the workpiece deformation to increase as lapping progresses. This will cause large deviation in surface form between finished workpiece and theoretical calculation. The mold with its surface approximates the desired surface form of the lens which is used for improving the machining precision. When vacuum pressure is supplied, the top surface of the workpiece will be deformed and then contacts the molded surface. **Figure 13** shows the basic concept of elastic deformation molding process [23].

The mold and workpiece surfaces are polished to flatness before fabricating as shown in **Figure 13a**. When uniform vacuum pressure is supplied to the workpiece through small holes of the mold, the workpiece will be deformed and then contacted with the aspheric surface of the mold as presented in **Figure 13b**. While deformed workpiece is kept stable under vacuum pressure, the bottom side of the workpiece is polished to flatness as illustrated in **Figure 13c** and **d**. Then, the vacuum pressure is not supplied; hence, the lapped side of workpiece will be formed into the mold surface while the opposite surface returns to its original flatness surface due to material elasticity as shown in **Figure 13e**.

4.1 Finite element analysis for elastic deformation machining process with mold

The standard aspheric formula is:

$$Z = \frac{cr^2}{1 + \sqrt{1 - (1+k)c^2r^2}} + \sum_{i=2}^{n} A_{2i}r^{2i} \tag{20}$$

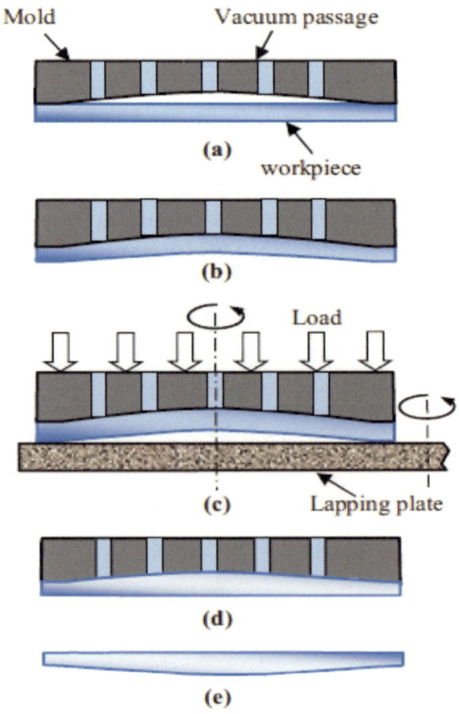

Figure 13.
Basic principle of elastic deformation molding process.

where Z, depth or "Sag" of the curve; r, distance from the center; c, curvature (=1/radius); K, conic constant; and A_{2i}, higher order terms.

The radius (R) is used for determining the aspheric terms such as their shallow or depth. The closest spherical surface is the radius which reaches the aspheric sag at the largest useful diameter [24]. **Figure 14** illustrates the aspheric lens sag.

An aspherical surface is built by using spherical surface combined with the higher order terms. Most optical designers use only the even-order terms from A2 to A20. The conic constant K has been used to design the initial aspheric, simple paraboloid and hyperboloid (as shown in **Table 3**).

In elastic deformation machining method, the accuracy of aspheric lens depends on the ability of elastic deformation and completely contacting the mold surfaces. The mold surface is defined by choosing the closest spherical surface (as shown in **Figure 15**). The FEA is designed for establishing the spherical surface through a simulation of contacting process between workpiece and mold surface.

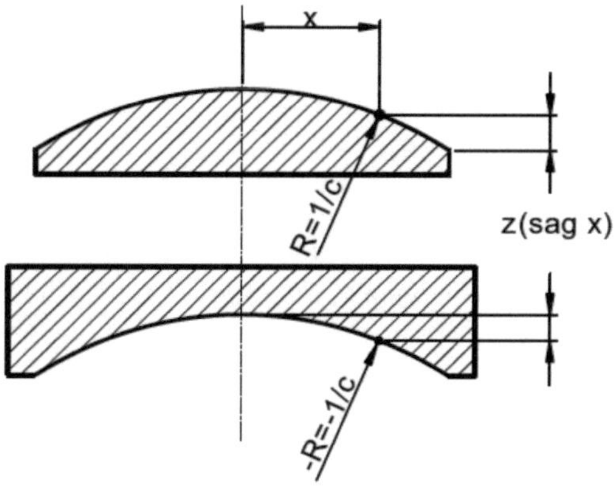

Figure 14.
The sag of aspheric lens.

Conic constant	Surface type
$K = 0$	Spherical
$K = -1$	Paraboloid
$K < -1$	Hyperboloid
$-1 < K < 0$	Ellipsoid
$K > 0$	Oblate ellipsoid

Table 3.
The relationship between conic constants and surface types.

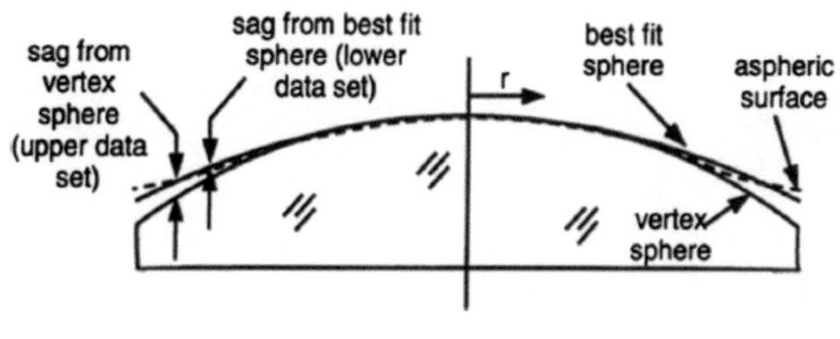

Figure 15.
The aspheric surface from best fit sphere.

In the simulation process, the thickness (h = 1.0 mm) and diameter (D = 50 mm) of the workpiece are suggested. In addition, the radius (R = 2500 mm) of spherical surface is chosen. The parameters of FEA model can be seen in **Figure 16a**.

The axisymmetric model is selected in this simulation process. The mold is chosen as an analytical rigid shell and the workpiece is a deformable shell. The analytical step of model is "Dynamic, Explicit". The interaction and the contact property are "Surface to surface contact" and "Penalty contact method," respectively. All elements of the workpiece are divided in meshing with A4-node bilinear axisymmetric quadrilateral elements in reduced integration. The workpiece mesh and boundary conditions are described in **Figure 16b**. The values of uniform vacuum pressure are opted in range of −80 to −100 kPa. The conic constant K = 0.25 is selected for the simulation process.

It is clear to see that **Figure 17** shows the deflection and deviation of workpiece under different vacuum pressures with the conic constant K = −3.

According to the results, the model with the conic constant K = −3, gives the best one and the deviation between the workpiece and the mold is the smallest. The workpiece and the mold can reach the best when the vacuum pressure approximates −95 kPa. However, when the vacuum pressure is larger than −95 kPa, the deviation results are still stable. Therefore, the conic constant K = −3 recommends for defining the aspheric surface of the mold.

The accuracy one can be innovated by modifying the mold profile to adopt with bending stress of workpiece material. This mold profile is redesigned by using the profile of workpiece after the deformed stage. An axisymmetric FEM model is established, and it consists of the new mold and workpiece. The uniform vacuum pressure is chosen as −95 kPa. The mold surface is redesigned with the conic constant K = −3 as displayed in **Figure 18**.

It can be noted that **Figure 19a** and **b** presents the deflection and deviation results between the workpiece and the new mold under supplied vacuum pressures, P = −95 kPa and K = −3.

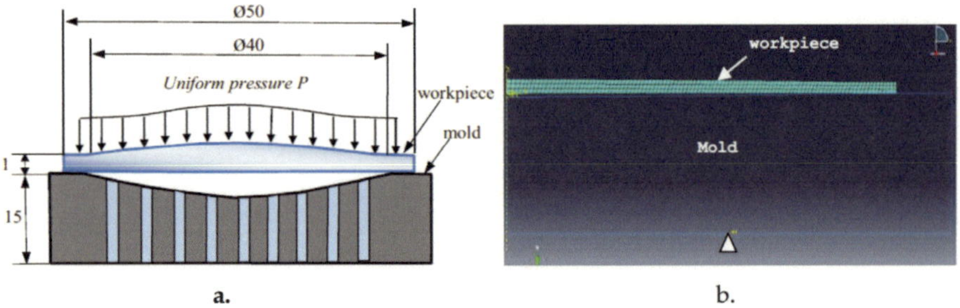

Figure 16.
(a) The simulation model and (b) FEM simulation model.

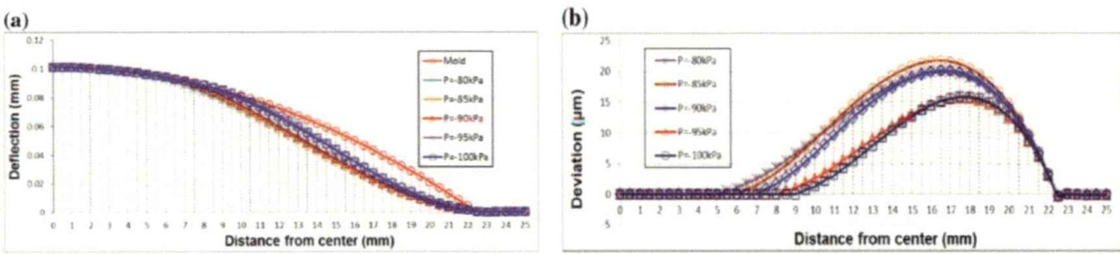

Figure 17.
(a and b) Deflection and deviation under different vacuum pressures (K = −3).

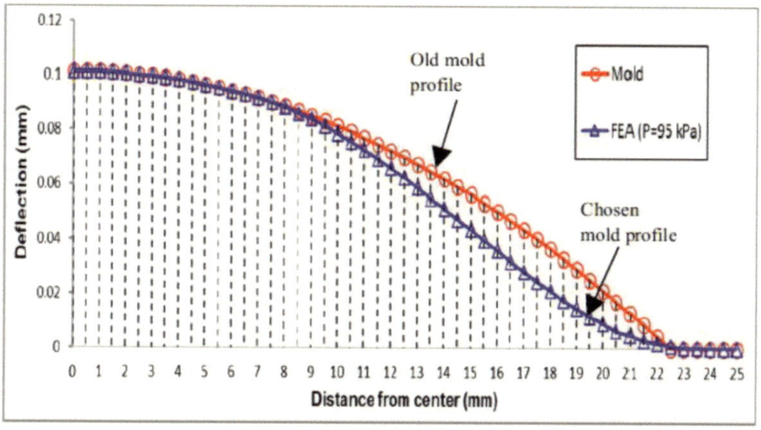

Figure 18.
The modified mold is chosen.

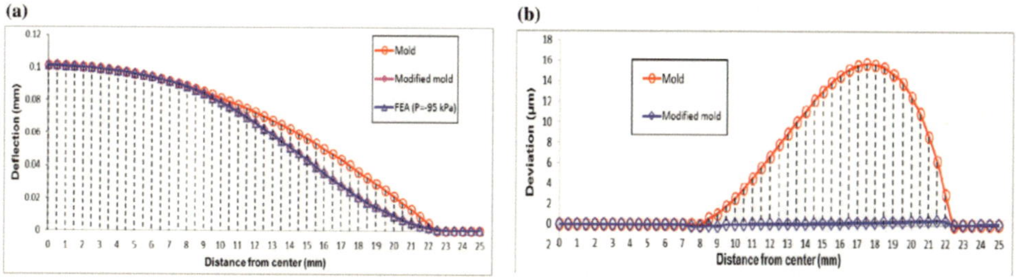

Figure 19.
(a and b) Deflection and deviation results between the workpiece and modified mold.

The form accuracy of workpiece is enhanced by using the new mold surface. The maximum deviation is less than *P-V* 0.35 μm while the former mold is about 15.02 μm.

4.2 Experimental setup

Figure 20 presents that the experiment was conducted to a precision polishing machine Preci-Polish 300. The B270 glass with a diameter of 50 mm and a thickness

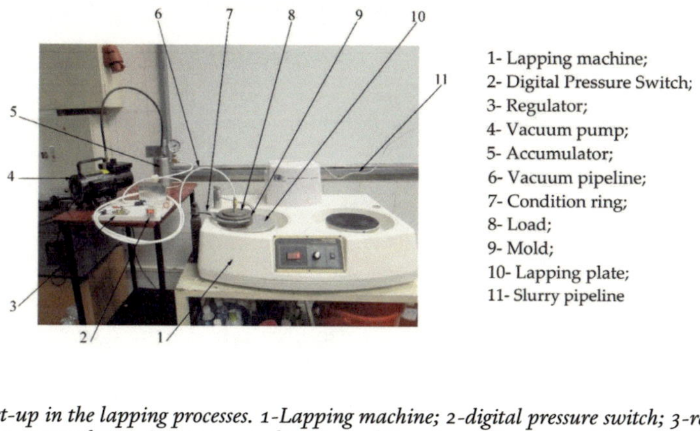

Figure 20.
Experimental set-up in the lapping processes. 1-Lapping machine; 2-digital pressure switch; 3-regulator; 4-vacuum pump. 5-accumulator; 6-vacuum pipeline; 7-condition ring; 8-load; 9-mold; 10-lapping plate; and 11-slurry pipeline.

of 1.0 mm is utilized in the experiment process. In addition, **Table 4** points the parameters for the machining process, in which the vacuum pressure is fixed as −95 kPa.

According to the simulation results, the new mold surface with the conic constant $K = -3$ is chosen. The multi-small holes of the modified mold surface are fabricated to fix the workpiece during the machining process. The new mold is machined on a precision CNC machining center.

4.3 Experimental results

Figure 21a and **b** shows that experimental results are compared to FEA with new mold surface under applied vacuum pressure $P = -95$ kPa.

Based on the experimental and FEA results, the deviation of workpiece is less than P-V 0.01 μm within the radius of about 12 mm. The maximum deviation is P-V 0.6 μm; however, the former mold is about 18.93 μm. It is clear to see that the experimental results agree greatly with FEA results. Therefore, the form accuracy of the workpiece is significantly improved when the new mold profile is redesigned according to the FEA results with $P = -95$ kPa and $K = -3$.

Items	Lapping	Polishing
Abrasive	#1000 SiC	#10,000 CeO_2
Abrasive concentration in slurry (wt%)	10%	10%
Machining load (N)	30	20
Rotating speed of lapping plate (rpm)	60	40
Machining time (min)	120	30

Table 4.
Lapping and polishing parameters.

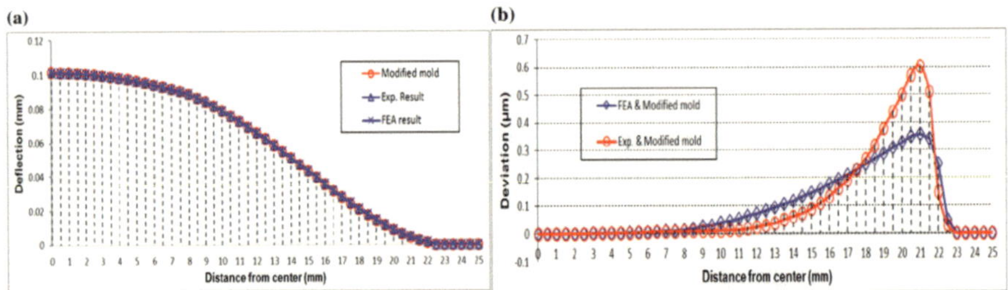

Figure 21.
(a and b) Experimental and FEA results with modified mold surface.

5. Conclusions

Based on the elasticity of the material, the elastic deformation machining is a method in which the vacuum pressure is used for fabricating complex aspheric surfaces. The amount of deflection of circular plate can be determined by solving the differential equations of an appropriate plate theory. The workpiece will be deformed into aspheric shape prior to the lapping process under the vacuum pressure. While the vacuum pressure is remained, the opposite side is polished to optical flatness by the lapping wheel. Then, the vacuum pressure is not supplied and hence, the bottom surface will be formed into the aspheric shape and the top surface will be restored to its flat surface form. Therefore, the method is suitable for manufacturing of optical lens with large aperture and low thickness glass materials.

In the elastic deformation machining process without mold, the manufactured workpiece accuracy can be increased by adjusting the vacuum pressure during the machining process because of the changed workpiece thickness. The vacuum pressure is defined through FEA results. According to the FEA, the deformation curve of the workpiece is reached to the desired curve when the vacuum pressure is fixed at 42 kPa. Depending on reducing the vacuum pressure from 50 to 42 kPa and keeping stable through the entire machining process, the experimental results agree greatly with theoretical calculations. The best peak-valley value *P-V* 1.6 μm was achieved in this method.

In order to achieve form accuracy of the workpiece in the elastic deformation machining process with mold, the mold with its surface approximates the desired surface form of the lens which is used for improving the machining precision. The accuracy one can be innovated by modifying the mold profile to adopt with bending stress of workpiece material. This mold profile is redesigned by using the profile of workpiece after the deformed stage. According to the simulation results, the new mold surface with the conic constant $K = -3$ and vacuum pressure $P = -95$ kPa are used for the experimental process. In this case, the deviation of workpiece is less than *P-V* 0.01 μm within the radius of about 12 mm. The maximum deviation is *P-V* 0.6 μm; however, the former mold is about 18.93 μm. It is clear to see that the experimental results agree greatly with FEA results.

Author details

Duc-Nam Nguyen
Faculty of Mechanical Engineering, Industrial University of Ho Chi Minh City, Vietnam

*Address all correspondence to: nguyennams@gmail.com

References

[1] Aono Y, Negishi M, Takano J. Development of large-aperture aspherical lens with glass molding. Advanced Optical Manufacturing and Testing Technology. 2000;**4231**:16-23. DOI: 10.1117/12.402759

[2] Suzuki H, Highuchi O, Huriuchi H, Shibutani H. Precision cutting of micro-axis-symmetric spherical surface with 3-axes controlled diamond tool. In: Proceedings of the Second Euspen International Conference on European Society for Precision Engineering and Nanotechnology; 27-31 May 2001; Italy. pp. 844-847

[3] Anurag J. Experimental study and numerical analysis of compression molding process for manufacturing precision aspherical glass lenses [thesis]. Columbus: Ohio State University; 2006

[4] Takashi M, Takenori O. Cutting process of glass with inclined ball end mill. Journal of Materials Processing Technology. 2008;**200**:356-363. DOI: 10.1016/j.jmatprotec.2007.08.067

[5] Suzuki H, Moriwaki T, Yamamoto Y, Goto Y. Precision cutting of aspherical ceramic molds with micro PCD milling tool. Annals of the CIRP. 2007;**56**: 131-134. DOI: 10.1016/j.cirp. 2007.05.033

[6] Kim HS, Kim EJ, Song BS. Diamond turning of large off-axis aspheric mirrors using a fast tool servo with on machine measurement. Journal of Materials Processing Technology. 2004; **146**:349-355. DOI: 10.1016/j. jmatprotec.2003.11.028

[7] Li L, Yi AY, Huang C, Grewell DA, Benatar A, Chen Y. Fabrication of diffractive optics by use of slow tool servo diamond turning process. Optical Engineering. 2006;**45**:113401. DOI: 10.1117/1.2387142

[8] Bifano TG, Dow TA, Scattergood RO. Ductile-regime grinding: A new technology for machining brittle materials. Journal of Engineering for Industry. Transactions of the ASME. 1991;**113**:184-189. DOI: 10.1016/ 0141-6359(92)90161-O

[9] Lambropoulus JC, Fang T, Funkenbusch PD, Jacobs SD, Cumbo MJ, Golini D. Surface micro-roughness of optical glasses under deterministic micro-grinding. Applied Optics. 1996; **35**:4448-4462. DOI: 10.1364/ AO.35.004448

[10] Namba Y, Abe M, Kobayashi A. Ultra-precision grinding of optical glasses to produce super-smooth surfaces. Annals of the CIRP. 1993;**42**: 417-420. DOI: 10.1016/S0007-8506(07) 62475-5

[11] Chen WK, Kuriyagawa T, Huang H, Yoshihara N. Machining of micro aspherical mould inserts. Journal of Precision Engineering. 2005;**29**:315-323. DOI: 10.1016/j.precisioneng.2004.11.002

[12] Brinksmeier E, Mutlugunes Y, Klocke F. Ultra-precision grinding. CIRP Annals-Manufacturing Technology. 2010;**59**:652-671. DOI: 10.1016/j.cirp.2010.05.001

[13] Bingham R, Walker D, Kim D. Novel automated process for aspheric surfaces. In: Proceedings of SPIE – The International Society for Optical Engineering. 2000;**4093**:445-448. DOI: 10.1117/12.405237

[14] Walker D, Brooks D, King A. The “Precessions” tooling for polishing and figuring flat, spherical and aspheric surfaces. Optics Express. 2003;**11**: 958-964:958-964. DOI: 1364/ OE.11.000958

[15] Nelson J, Scordato M, Schwertz K. Precision lens molding of asphero

diffractive surfaces in chalcogenide materials. Optifab. 2015:96331L. DOI: 10.1117/12.2195764

[16] Mahajan P, Dora PT, Sandeep TS. Optimized design of optical surface of the mold in precision glass molding using the deviation approach. International Journal for Computational Methods in Engineering Science and Mechanics. 2015;**16**:53-64. DOI: 10.1080/15502287.2014.976677

[17] Yi AY, Tao B, Klocke F. Residual stresses in glass after molding and its influence on optical properties. Procedia Engineering. 2011;**19**:402-406. DOI: 10.1016/j.proeng.2011.11.132

[18] Kunz A. Aspheric freedoms of glass-precision glass moulding allows cost-effective fabrication of glass aspheres. Optik & Photonik. 2009;**4**:46-48. DOI: 10.1002/opph.201190063

[19] Mori Y, Yamamura K, Endo K, Yamauchi K, Yasutake K, Goto H, Kakiuchi H, Sano Y, Mimura H. Creation of perfect surfaces. Journal of Crystal Growth. 2005;**275**:39-50. DOI: 10.1016/j.jcrysgro.2004.10.097

[20] Ventsel E, Krauthammer T, editors. Thin Plates and Shells: Theory, Analysis, and Applications. 1st ed. Boca Raton: CRC Press; 2001. p. 688

[21] Nguyen DN, Lv BH, Yuan JL, Wu Z, Lu HZ. Experimental study on elastic deformation machining process for aspheric surface glass. International Journal of Advanced Manufacturing Technology. 2013;**65**:525-531. DOI: 10.1007/s00170-012-4191-3

[22] Schott. Schott B270 Super-white Properties [Internet]. 2015. Available from: https://psec.uchicago.edu/glass/SchottB270Properties-KnightOptical.pdf [Accessed: 24-05-2018]

[23] Nguyen DN. Study on improving the precision of form surface produced in elastic deformation molding process. International Journal of Advanced Manufacturing Technology. 2017;**93**:3473-3484. DOI: 10.1007/s00170-017-0766-3

[24] Kweon G, Kim CH. Aspherical lens design by using a numerical analysis. Journal of the Korean Physical Society. 2007;**51**:93-103. DOI: 10.3938/jkps.51.93

Chapter 4

Concept of Phase Transition Based on Elastic Systematics

Paul S. Nnamchi and Camillus S. Obayi

Abstract

The use of elastic constants systematics to describe fundamental properties of engineering materials has made materials science education and its related subjects increasingly important not only for manufacturing engineers but also for mankind at large. In this chapter, we present actual scaling of phase transition-driven considerations, such as martensitic transformation and transformable shape memory formation via elastic constant systematics. The scaling in terms of the simple and polycrystals mechanical stability criteria based on the elastic moduli and an acoustic anisotropy is in good agreement with novel experimental data from the literatures, and further, a long-standing concern in predicting polycrystalline elastic constants was considered beyond the commonly encountered criteria.

Keywords: elastic, elastic modulus, martensitic transformation, shape memory effect, elastic constant, ductility criterion, mechanical properties

1. Introduction

The ingenuity and the art required to tailor precisely the desired physical and structural properties in materials have been the main goal of the material scientists and engineers. Elastic response (i.e. elastic constant) to an applied load is one of such basic properties of all solids and originates from the distortion of atomic bonds. Simply put, elastic constants are a reflection of the fundamental thermodynamic properties that take place in the crystal lattice of solids. Complementary to this, the otherwise inaccessible essential information can be revealed from their temperature and stress dependencies of these important constants. For instance, the crystal structures of the three long periods of transition elements change more or less systematically from hcp through bcc to fcc as their group numbers increase from IV to VIII as does their elastic properties. Thus, the knowledge of microscopic elasticity can provide a fruitful ground for the exploration of the material behaviour yet uncommon to our knowledge about the relationship between crystal structure and bonding.

The earliest foundation of elastic theory dates back to seventeenth century (around 1821), when Navier first gave the equation for the equilibrium and motion of elastic solids [1], but modern foundation of microscopic elastic theory was established by the work of Born and Huang [2], followed by other excellent treatments [3]. It is well known that crystalline solids are by no means ideal and invariably contain some lattice defects such as vacancies, solute atoms or some extent of disorder. These point defects strongly affect almost all properties of materials, including elastic behaviours. In effect, the early investigators of these phenomena were motivated by the response of naturally occurring anisotropic materials such as wood and other crystalline solids. On that premise, of interest here is the relationship between crystal structure and elastic properties, mainly because of the important information they provide about nature of binding forces in solids.

Over the past three decades, elastic constants of some simple crystals have been a subject of numerous researches and have been investigated both theoretically and experimentally. Some of the outcomes have revealed that fundamental elastic properties of a martensitic crystal are fully determined by the elastic constants C_{ij}. All macroscopic elastic moduli (Young's and shear modulus, Poisson ratio, etc.) can be derived from the C_{ij} at least within certain upper and lower bounds [4]. There is considerable evidence that the magnitude of $C^{'} = (C_{11} - C_{12})/2$ elastic shear modulus in metallic bcc structures is closely related to the occurrence of martensitic phase transformations and is thus a useful parameter for estimating bcc structures [5]. Similarly, whether a structural material shows plastic flow or brittle fracture on loading is of clear practical significance. Brittleness in polycrystalline metals can be intrinsic or induced. The basic question is: Do these two general properties (i.e. phase stability and elastic properties) of crystals correlate to each other?

2. Analytical criterion of elastic constants of perfect crystals

The elastic properties are among the most important physical properties of materials and the importance of studying elastic properties of materials cannot be overemphasised. The knowledge of elastic properties is essential for both structural design and experimental mechanics [6]. It also enables the assessment of the sufficiency of strength, stiffness and stability of newly developed materials. Although the crystals are assumed to free from lattice imperfections and difficult to produce, their study had always been the building block for a better understanding of the behaviour of bulk materials. Usually, the determination of elastic properties of crystalline solids is based on its single or perfect crystal configuration under special loading conditions. The elastic moduli are the material constants that connect stress with strain and are therefore crucial to engineering applications. A crystal subjected to external load undergoes dimensional change. If the eternal load is a stress tensor denoted by σ_{ij}, then the deformation per unit length in three-dimensional space, can be described by a strain tensor, e_{ij}. Within the elastic limit or for sufficiently small deformations, the stress tensor is a linear function of the strain tensor and the generalised delta notation of Hooke's law can be used to express the relationship between these two quantities [7] as:

$$\sigma_{ij} = C_{ijkl}e_{kl} \tag{1}$$

where C_{ijkl} is the proportionality constant that characterises the crystal's resistance to elastic shape change; often referred to as the elastic coefficients or elastic constants or elastic moduli or stress-strain coefficients [8].

The inverse relation between the strain and the stress can be determined by taking the inverse of stress-strain relation to get:

$$e_{ij} = S_{ijkl}\sigma_{kl} \tag{2}$$

Here, S_{ijkl} represents the elastic compliance of the crystal. From symmetry or equilibrium principles, the state of stress in an elastic body can be approximated by six independent stress and strain components. And as such the stress and strain components in Eq. (1) can be expressed in three orthogonal axes as:

$$\begin{aligned}
\sigma_x &= C_{11}e_{xx} + C_{12}e_{yy} + C_{13}e_{zz} + C_{14}e_{yz} + C_{15}e_{zx} + C_{16}e_{xy} \\
\sigma_y &= C_{21}e_{xx} + C_{22}e_{yy} + C_{23}e_{zz} + C_{24}e_{yz} + C_{25}e_{zx} + C_{26}e_{xy} \\
\sigma_z &= C_{31}e_{xx} + C_{32}e_{yy} + C_{33}e_{zz} + C_{34}e_{yz} + C_{35}e_{zx} + C_{36}e_{xy} \\
\tau_{yz} &= C_{41}e_{xx} + C_{42}e_{yy} + C_{43}e_{zz} + C_{44}e_{yz} + C_{45}e_{zx} + C_{46}e_{xy} \\
\tau_{zx} &= C_{51}e_{xx} + C_{52}e_{yy} + C_{53}e_{zz} + C_{54}e_{yz} + C_{55}e_{zx} + C_{56}e_{xy} \\
\tau_{xy} &= C_{61}e_{xx} + C_{62}e_{yy} + C_{63}e_{zz} + C_{64}e_{yz} + C_{65}e_{zx} + C_{66}e_{xy}
\end{aligned} \tag{3}$$

Here, e_{xx}, e_{yy} and e_{zz} are tensile strains, e_{xy}, e_{yz} and e_{zx} are shear strains. The experimental values of elastic constants, C_{ijkl}, were originally determined by considering the response of crystals to small strains or unstressed lattice using Eq. (1). Beyond using Eq. (1) based on measured stress-strain relations, there are now methods of determining elastic constants from the first principles often referred to as ab initio methods. There are many methods of evaluating elastic coefficients such as the one based on expanding the internal strain energy of the crystal [7]. Thus, we may write as Eq. (4),

$$U = U_0 + V_0 \sum \sigma_i e_i + \frac{1}{2} V \sum_i \sum_i C_{ij} e_i e_j + \dots \tag{4}$$

where U is the energy of the crystal, is a quadratic function of the strains, in the approximation of Hooke's law (recall the expression for the energy of a stretched spring). V_0 is its equilibrium volume and e denotes an elastic strain. If the material is a crystal, the number of independent elastic constants is reduced further depending on the crystal system.

Elastic coefficients and elastic moduli have significant effect of mechanical response of crystals. Elastic constants, $C_{ij}(C_{11}, C_{12}, C_{44})$ and elastic moduli such as bulk modulus (B), shear modulus (G), Young's modulus (E) influence mechanical response of crystals. For instance, the bulk modulus (B) is associated with the hardness of materials which is of extreme importance in high-temperature and pressure applications, while elastic constants could provide essential information about bonding between adjacent atomic planes, anisotropic character of bonding and structural stability [7]. By far, the most widely reported elastic properties are E, G and B, corresponding to tensile, shear and hydrostatic loading, respectively. Since B signifies the compressibility of a substance, it can be calculated from the partial derivative of volume (V) and pressure (P) at constant temperature (T), as per Eq. (5).

$$B = \left({}^{\delta V}\!/_{\delta P} \right)_T \tag{5}$$

It is worth pointing out that other definitions of elastic constant are possible. Elastic modulus *E* measures the resistance to a change in atomic separation distance within the plane of the bond and so can be determined from the linear portion of the interatomic potential. *G* quantifies the resistance to shear loading and *B*, since it corresponds to a volumetric dilatation, is dependent on the electronic properties of a solid, i.e. the compressibility of the electron gas. Elastic moduli are therefore controlled by interatomic interactions and so may be considered a fundamental property of condensed matter. By excitation of longitudinal and transverse phone modes, E and G can, respectively, be calculated if the density (ρ) of the material is known. This is done via an ultrasonic probe which emits and measures the longitudinal (v_l) and transverse (v_t) sound wave velocities, from which E and G can be calculated via Eqs. (6) and (7):

$$E = \rho v_l^2 \tag{6}$$

$$G = \rho v_t^2 \tag{7}$$

E, *G* and *B* can also be calculated from C_{ij} elastic constants. For a material with cubic structure, the number of C_{ij} in the elastic tensor can be reduced from 36 to just 9, due to $C_{ij} = C_{ji}$ and there being strong symmetry in a cubic lattice. The resulting relevant C_{ij} are C_{11}, C_{12} and C_{44}.

$$C_{12} = B + \frac{4G}{3} \tag{8}$$

$$C_{11} = 3B - \frac{C_{11}}{2} \tag{9}$$

$$C_{11} = G \tag{10}$$

$$C' = C_{11} - \frac{C_{12}}{2} \tag{11}$$

The tetragonal shear modulus, *C*, corresponds to a specific phonon vibration mode in the atomic structure, and is thus directional in nature. In comparison, B is non-directional as it relates to a volumetric effect.

$$B = \frac{C_{11} + 2C_{12}}{3} \tag{12}$$

$$G = \frac{C_{11} - C_{12}}{3} \tag{13}$$

3. Elastic and lattice stability criteria

3.1 Lattice stability in perfect crystal

Elastic properties of a material are very important because they check the mechanical stability, ductile or brittle behaviour based on the analysis of elastic constants, C_{ij}, bulk modulus B and shear modulus G. For example, the bulk modulus measures the resistance of the volume variation in a solid and provides an estimation of the elastic response of the materials under hydrostatic pressure. The shear modulus describes the resistance of a material to shape change.

The fundamental understanding of the conditions of mechanical stability of unstressed crystal structure was laid by the work of Max-Born and co-authors in the 1940s [3], and consolidated later in 1954 [3]. This and other text books gave the generic requirements for elastic stability of crystal lattices in terms of elastic constants [3] and offers simplified equivalents of the generic conditions for some high-symmetry classes. The general stability condition can be stated by considering the second-order elastic matrix and the elastic energy of the crystal deformed homogeneously by infinitesimal strain as shown in Eqs. (14) and (15) [3], respectively:

$$C_{ij} = \frac{1}{V_0}\left(\frac{\partial^2 U}{\partial^2 e_i \partial^2 e_j}\right) \tag{14}$$

$$U = \frac{1}{2} V_0 \sum_{i,j=1}^{6} C_{ij} e_i e_j + O(e^3) \tag{15}$$

where U is the elastic energy, V_O is the volume of unstressed sample, C_{ij} (I, j = 1–6) is the elastic constant and e_i and e_j are the applied strains [2]. In Eq. (15), O (e3) denotes the terms of numerical error in the order e^3 or higher. A crystal lattice is dynamically said to be stable only if elastic energy U is positive for any small deformation [9], which implies that principal minors of the determinant with elements Cij are all positive [3].

Most real materials (cubic and non-cubic polycrystalline structures) have some types of symmetry, which further reduces the required number of independent elastic moduli. In the case of cubic systems, such as bcc, fcc, NaCl type, or CsCl type) structures, in particular, number of independent elastic moduli is reduced from 36 to 9, as C_{ij} = C_{ji} and there being strong symmetry in the two lattices. Therefore, the conditions for stability reduced to a very simple form using three different elastic constants: C_{11}, C_{22} and C_{44}. The mechanical stability criteria are given by [10]:

$$\begin{gathered} C_{11} - |C_{12},|>0 \\ C_{11} + 2\,C_{12}>0 \\ C_{44}>0 \\ C_{12}>C_{11} \end{gathered} \tag{16}$$

The condition when $B < 0$ is referred to as spinodal instability.

Although hexagonal and tetragonal systems have the same form for the elastic matrix, the hexagonal has five, while tetragonal has six independent elastic constants. By direct calculation of the Eigen values of the stiffness matrix, according to [11], four conditions can be derived for elastic stability in both classes:

$$\begin{gathered} C_{11}>|C_{12},|;2C_{13}^2 < C_{33}(\,C_{11} + C_{12}) \\ C_{44}>0;C_{66}>0 \end{gathered} \tag{17}$$

Similarly, for the orthorhombic system, there are nine independent elastic constants: C_{11}, C_{22}, C_{33}, C_{44}, $C_{12}C_{55}$, C_{66}, C_{23} and C_{13}. The mechanical stability of the structure at each concentration can be judged by calculated elastic stiffness. According to Born's criteria [3], the requirement of mechanical stability in an orthorhombic system leads to the following equations [12].

$$\begin{gathered} C_{11}>0;C_{22}>0;C_{44}>0;C_{33}>0;C_{55}>0;C_{66}>0; \\ C_{11} + C_{22}>2\,C_{12};C_{11} + C_{33}>2\,C_{13}; \\ C_{11} + C_{22}\,C_{33} + 2\,C_{12} + 2\,C_{23} + 2\,C_{13}>0; \end{gathered} \tag{18}$$

Further to this, conditions for stability for some high-symmetry crystal classes have been studied. However, there is still some confusion about the form of stability criteria for other crystal systems and classes [8].

A crystal lattice is said to be stable in the absence of external load (unstressed condition) and in the harmonic approximation [13] if and only if it has both dynamic and elastic stability. Dynamic stability implies that its phonon modes have positive frequencies for all wave vectors, while its elastic stability is dependent on elastic energy given by Eq. (15) being always positive ($U > 0, \forall \varepsilon \neq 0$). Elastic stability criterion is mathematically equivalent to the following necessary and sufficient conditions: the elastic matrix C is definite exactly positive and all Eigen values of matrix C are positive; all the leading principal and arbitrary minors of matrix C are all positive. The closed form expressions for necessary and sufficient elastic stability criteria for other crystal lattices have been studied. While the stability criterion is linear for some crystal lattices, it is quadratic and even polynomial for others. Thus, the mechanical stability of a crystal is combination of the elastic constant and Born's stability criteria. The elastic constant of a stable crystal must satisfy the Born's criteria to prove its mechanical stability.

3.2 Relative stability of polycrystalline materials

In the case of multi-phase stability, multi-phase composites can be obtained based on multiple scattering theory. For example, polycrystalline materials consisting of two phases, namely cubic and orthorhombic phases can be obtained by homogenising the integral elastic response of the multi-phase polycrystalline samples, following the effective medium approach originally applied by Zeller and Dederichs [13] to determine elastic properties of single-phase polycrystals with cubic symmetry. This type of concept was generalised by Middya and Basu [14] and further extended by Middya [15] and by Raabe et al. [16] to multi-phase composites to determine: (i) the elastic single constants and (ii) the volume fraction of the components within a self-consistent *T*-matrix solution for the effective medium elastic properties of hexagonal, and orthorhombic polycrystals.

The subset of supercells or cubic and orthorhombic symmetries consisting of three (C_{11}, C_{12}, C_{44}) and nine (C_{11}, C_{12}, C_{13}, C_{22}, C_{23}, C_{33}, C_{44}, C_{55}, C_{66}) elastic constants, respectively, was calculated by employing the methodology explained in [16–18] for the elastic properties of the multi-component alloys. This can be viewed as a macroscopic homogeneous effective medium consisting of microscopic fluctuations and characterised by an effective stiffness of C_{ijkl} defined by:

$$\langle \sigma_{ij}(\boldsymbol{r}) \rangle = C_{ijkl} \langle \epsilon_{kl}(\boldsymbol{r}) \rangle \tag{19}$$

Here, C_{ijkl} is the local elastic constant tensor with $\langle \sigma_{ij}(\boldsymbol{r}) \rangle$ and $\langle \epsilon_{kl}(\boldsymbol{r}) \rangle$ as the local stress and strain field at a point **r**, respectively, and the angular brackets denoting ensemble averages. A repeated index implies the usual summation convention. The effective stiffness of C_{ijkl} is defined by:

$$\langle \sigma_{ij}(\boldsymbol{r}) \rangle = C^{*}_{ijkl} \langle \epsilon_{kl}(\boldsymbol{r}) \rangle \tag{20}$$

Since the aggregate represents a body in equilibrium, $\sigma_{ij}(r)|j = 0$, where $|j = \partial / \partial r_j$ and the local elastic constant tensor can now be decomposed into an arbitrary constants part (C^{o}_{ijkl}) and a fluctuating part—$\delta C(\boldsymbol{r})$.

$$C_{ijkl}(r) = C^{o}_{ijkl} + \delta C_{ijkl}(r) \tag{21}$$

As shown in [16], an integral part of Eq. (19) is the interactive equivalent solution representing the resulting local strain ϵ distribution (in a short notation) as:

$$\epsilon = \epsilon^0 + GT\epsilon^0, \tag{22}$$

Here, ϵ^0 and G are the strain and modified Green's function of the medium defined by C^O, and the T-matrix given by:

$$T = \delta C(1 - G\delta C)^{-1} \tag{23}$$

Here, $\boldsymbol{I}$ is equivalent to the unit tensor. Combining Eqs. (21) and (22), we get:

$$C^* = C^0 + \langle T\rangle/(1 + \langle GT\rangle)^{-1} \tag{24}$$

Although Eq. (21) constitutes an exact solution for C^*, finding the exact solution of $\langle T\rangle$ and $\langle GT\rangle$ for realistic cases is impossible. By neglecting the intergranular scattering that may occur in some cases in the form of a grain-to-grain position-orientation correlation function however, the T-matrix can be written in terms of single-grain T-matrix (t_α) for each grain α

$$T \approx \sum_\alpha t_\alpha = \tau. \tag{25}$$

where

$$t_\alpha = \delta C_\alpha + \delta C_\alpha G t_\alpha = \delta C_\alpha (1 - G\delta C_\alpha)^{-1} \tag{26}$$

$$\sum_\alpha \delta C_\alpha = \delta C = C - C^0 \tag{27}$$

Inserting Eq. (21) into (22) leads to:

$$C^* = C^0 + \langle\tau\rangle(1 + \langle G\tau\rangle)^{-1} \tag{28}$$

For single-phase polycrystal, the self-consistent solution of Eq. (11) can be obtained by choosing a C^* that satisfies:

$$\langle\tau\rangle = 0 \tag{29}$$

For a multi-phase polycrystals, a solution to Eq. (4) can be found by evaluating the volume fraction and τ of each phase $i(u^2 \ and \ \tau^2)$, respectively [19], via:

$$\left\langle \sum_i v^2\tau^2 \right\rangle = 0 \tag{30}$$

The application of the method to both single-phase aggregates and multi-phase composites is relevant to many multi-component alloys. For a single-phase polycrystal with cubic symmetry [16, 20] to the following expression for B^* and $\mu^* : B^* = B^o$

$$8\mu^{*3} + \left(9B_0 + 4C^{''}\right)\mu^{*2} - 3C_{44}\left(B_O + 4C^{''}\right)\mu^* - 6B_O C_{44} C^{''} = 0 \tag{31}$$

In Eq. (31), three independent single-crystal elastic constants (C_{11}, C_{12}, C_{44}) define the single-crystal bulk modulus $B^o = \left(C^0_{11} + 2C^O_{12}\right)/3$, the tetragonal shear modulus $C' = (C_{11} - C_{12})/2$ and trigonal shear modulus, C_{44}, $\mu^* = C^*_{44} = \mu^0 + \frac{\langle\tau_{44}\rangle}{1+G_{44}\tau_{44}}$.

Here, C_{44}^{*} is the homogenised bulk modulus. The details of the equation for calculating the elastic constants of polycrystals alloy with hexagonal symmetry have been explained elsewhere by [20], and the details here concern polycrystals with orthorhombic symmetry. Eqs. (29) and (30) are reduced to a set of coupled equations for B^{*} and μ^{*}:

$$0 = 9(K_V - B^{*}) + 2\beta(d - c + e) + 3\beta^2\Delta' \tag{32}$$

$$0 = \frac{a - b + \beta(2d - 2c - e) + 3\gamma(d - c + e) + \eta\beta\Delta'}{1 - \alpha\beta - 9\gamma(k_v - B_0) + \beta(\beta + 2\gamma)(c - d) - 2e\beta\gamma - \frac{1}{3}\eta\beta^2\Delta''} + 3\left(\frac{C_{44} - \mu^O}{1 - 2\beta(C_{44} - \mu^O)} + \frac{C_{55} - \mu^O}{1 - 2\beta(C_{55} - \mu^O)} + \frac{C_{66} - \mu^O}{1 - 2\beta(C_{66} - \mu^O)}\right) \tag{33}$$

where

$$9K_v = C_{11} + C_{22} + C_{33} + 2(C_{12} + C_{13} + C_{23}), \tag{34}$$

$$B = ⅓(C_{11} + 2C_{12})\mu^{*} = C_{44} \tag{35}$$

$$\gamma = ⅑(\eta - 3\beta) \tag{36}$$

$$a = \delta C_{11} + \delta C_{22} + \delta C_{33}; b = \delta C_{12} + \delta C_{13} + \delta C_{23} \tag{37}$$

$$c = \delta C_{11}\delta C_{22} + \delta C_{11}\delta C_{33} + \delta C_{22}\delta C_{33}; d = \delta C_{12}^2 + \delta C_{13}^2 + \delta C_{23}^2 \tag{38}$$

$$e = \delta C_{12}\delta C_{13} + \delta C_{12}\delta C_{23} + \delta C_{13}\delta C_{23} - \delta C_{11}\delta C_{23} - \delta C_{22}\delta C_{13} - \delta C_{33}\delta C_{12} \tag{39}$$

$$\Delta' = \delta C_{11}\delta C_{22}\delta C_{33} + 2\delta C_{12}\delta C_{13}\delta C_{23} - \delta C_{11}\delta C_{23}^2 - \delta C_{22}\delta C_{13}^2 - \delta C_{33}\delta C_{12}^2 \tag{40}$$

$$\delta C_{11} = C_{11} - C_{11}^O = C_{11} - K^0 - \frac{4}{3}\mu^0; \delta C_{22} = C_{22} = C_{22} - K^0 - \frac{4}{3}\mu^0 \tag{41}$$

$$\delta C_{33} = C_{33} - K^0 - \frac{4}{3}\mu^0; \delta C_{12} = C_{12} - C_{12}^O = C_{12} - K^0 + \frac{4}{3}\mu^0 \tag{42}$$

$$\delta C_{13} = C_{13} - K^0 + \frac{4}{3}\mu^0; \delta C_{23} = C_{23} - K^0 + \frac{4}{3}\mu^0 \tag{43}$$

$$\beta = \frac{-3(B^{*} + 2\mu^{*})}{5\mu^{*}(3B^{*} + 4\mu^{*})}, \tag{44}$$

$$\eta/3 = -1/3B^{*} + 4\mu^{*}, \tag{45}$$

$$C_{66} = (1/2)(C_{11} - C_{12}) \tag{46}$$

and orthorhombic symmetry has nine of the single crystal elastic constants, namely: C_{11}, C_{22}, C_{33}, $C_{44}C_{55}$, C_{66}, C_{12}, C_{23} and C_{13}.

The elastic constants of a multi-phase polycrystals were determined directly by coupling Eq. (13) for τ_{44} and the $(\tau_{11} + 2\tau_{12})$ components of the T-matrix. For materials with cubic symmetry, the equation is defined as:

$$5\tau_{44} = \left(\frac{1}{C_{11} - C_{12} - \widetilde{G}^{*}}^{-\beta}\right)^{-1} + 3\left(\frac{1}{C_{44} - \widetilde{G}^{*}}^{-2\beta}\right)^{-1} \tag{47}$$

$$\tau_{11} + 2\tau_{12} = \frac{3(C_{11} + 2C_{12}) - 9\widetilde{B}^{*}}{3 - (C_{11} + 2C_{12}) - 3\widetilde{B}^{*}} \tag{48}$$

This is where β is defined in Eq. (31) with $\widetilde{G}^{*}$ and $\widetilde{B}^{*}$ replacing G^{*} and B^{*}. For materials with orthorhombic symmetry, the equation reads:

$$15\tau_{44} = \frac{a - b + \beta(2d - 2c - e) + 3\gamma(d - c + e) + \upsilon\beta\Delta'}{1 - \alpha\beta - 9\gamma\left(K_\upsilon - \widetilde{B}_0\right) + \beta(\beta + 2\gamma)(c - d) - 2e\beta\gamma - \frac{1}{3}\upsilon\beta^2\Delta''}$$
$$+3\left(\frac{C_{44} - \widetilde{G}^O}{1 - 2k\left(C_{44} - \widetilde{G}^O\right)} + \frac{C_{55} - \widetilde{G}^O}{1 - 2k\left(C_{55} - \widetilde{G}^O\right)} + \frac{C_{66} - \widetilde{\mu}^O}{1 - 2\beta\left(C_{66} - \widetilde{G}^O\right)}\right) \quad (49)$$

$$\tau_{11} + 2\tau_{12} = \frac{9\left(K_\upsilon - \widetilde{B}^0\right) + 2\beta(d - c + e) + 3\beta^2\Delta'}{3\left[1 - \alpha\beta - 9\gamma\left(K_V - \widetilde{B}^0\right) + \beta(\beta + 2\gamma)(c - d) - 2e\beta\gamma - \frac{1\eta\beta^2\Delta'}{3}\right]} \quad (50)$$

Here, β is defined in Eq. (28), η is defined in Eq. (29), and Δ' in Eq. (23). Here, again G^* and $\widetilde{B}^*$ replaces G^* and B^* in the equations for β, υ and Δ'. As soon as G^* and $\widetilde{B}^*$ have been determined, the homogenised Young's modulus $\left(\widetilde{E}\right)^*$ and Poisson's ratio $(\upsilon)^*$ for (an elastically isotropic) polycrystal can be determined using standard elasticity relationships. The homogenised polycrystalline Young's modulus is calculated using:

$$E^* = \frac{9\widetilde{B}^* G^*}{3\widetilde{B}^* + G^*} \quad (51)$$

$$G^* = \frac{3\widetilde{E}^*\widetilde{B}^*}{9\widetilde{B}^* - \widetilde{E}^*} \quad (52)$$

4. Correlation of elastic constants with properties of polycrystalline materials

In many problem relating to polycrystalline or anisotropic materials, it is customary to make use of the properties in an elastically isotropic materials. Most of the common metals and engineering alloys, however, exhibit a marked degree of anisotropy in their single-crystal elastic behaviour and it is therefore more desirable to obtain same on the bases of anisotropic elastic property. The fundamental factors determining the intrinsic plasticity or brittleness behaviour in solids have great link with interatomic potentials, for instance, there is a correlation with the ratio of the elastic shear modulus μ to the bulk modulus B. It is evident, elastic moduli show trends with a range of properties, including hardness, yield strength, toughness and fragility [21, 22]. In this section, for limitation of space, we will, in particular, consider elastic aspect of polycrystals materials with respect to their dependency on specific crystal structure.

4.1 Elasticity and ductility criteria

Strength and ductility have always been one of the crucial issues to study for metal materials. The tendency of materials to be ductile or brittle is being predicted using models based on elastic constants. Some of these include that of Pugh criterion [23] and Cauchy pressure as defined by Pettifor [24]. Pugh proposed an empirical relationship between the plasticity and fracture properties showing the ratio G/B indicates the intrinsic ability of a crystalline metal to resist fracture and deform plastically [25]. This represents a competition between plasticity and fracture considering that B and G represent resistance to fracture and plastic deformation, respectively. Thus, the force required to propagate a dislocation is proportional to Gb where b is the Burgers vector. This implies that a material with high value of the ratio tends to be brittle (fracture is easier and plasticity is much less), while a low value indicates ductility (plasticity is easier and fracture is not). Fracture strength is also proportional to Ba (a, is the lattice constant) since B is related to surface energy, which indicates brittle fracture strength.

These empirical observations implicate G/B as explaining well brittle or tough behaviour [19, 26]. Pugh's criterion is the most widely used model to predict plastic behaviour of materials [27]. Since yield strength and fracture stress scale with shear modulus and elastic constant, respectively, the Pugh's ratio determines the likelihood of material's failure. If the effect of crystal structure is neglected, high value of Pugh's ratio indicates that a material is prone to brittle failure, while low value of G/B implies ductile failure. The large data on polycrystalline pure metals collected by Pugh [2], when he provided a qualitative ranking from ductile (e.g. Ag, Au, Cd, Cu) to brittle (e.g. Be, Ir) behaviour as G/B increases. For cubic close-packed (ccp) metals, the critical ratio $(G/B)_{crit}$ dividing the two regimes is in the range 0.43–0.56, and for hexagonal close-packed metals, it is 0.60–0.63. Cottrell [28] has estimated $(G/B)_{crit}$ for transgranular fracture from measured surface energies: 0.32–0.57 for ccp metals and 0.35–0.68 for body-centred cubic metals. The spread in values for each structure type largely indicates the interrelationship between crystal structure and elastic constant. Each structure type, however, includes metals with widely differing degrees of elastic anisotropy. Detailed analysis requires knowledge of the relevant elastic constants.

On the other hand, the Cauchy pressure ductility criterion is associated with elastic constants of single cubic crystals such as C_{12}–C_{44} and is useful in describing the nature of bonding in a material [27]. When a material has high resistance to bond bending as found in covalently bonded solids, it will have a negative Cauchy pressure ($C_{44} > C_{12}$). This is in contrast with materials with metallic bonding which exhibit positive Cauchy pressure. When compared with Pugh's ductility criterion, ductile and brittle behaviours are considered to be indicated by a positive and a negative Cauchy pressure, respectively. Although Pugh's and Cauchy pressure criteria are adjured to be based on easily measurable properties of materials such as elastic constants, they do not give the critical value dividing brittle and ductile materials. It is proven in certain materials, including metallic glasses and composites, which religiously respect this dividing line [21]. The behaviour is shown graphically in **Figure 1**. A summary of the correlation between C_{12}–C_{44} and $(G/B)_{crit}$

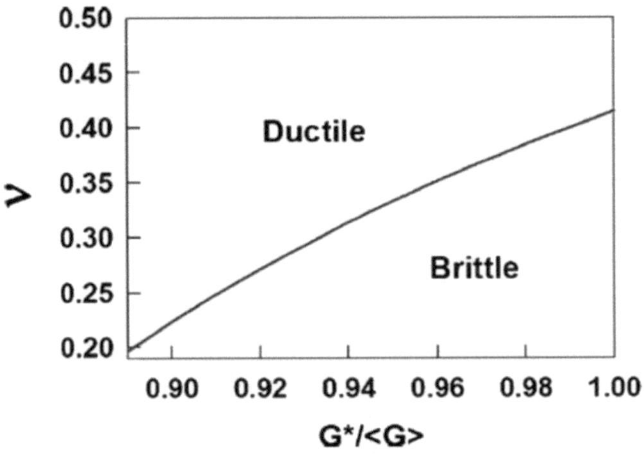

Figure 1.
Ductile and brittle phase fields in metallic glasses, where G is the local modulus and 'G' the global modulus. Decreasing the fraction of low G sites reduces the need for a globally low* ν *(culled from [28]).*

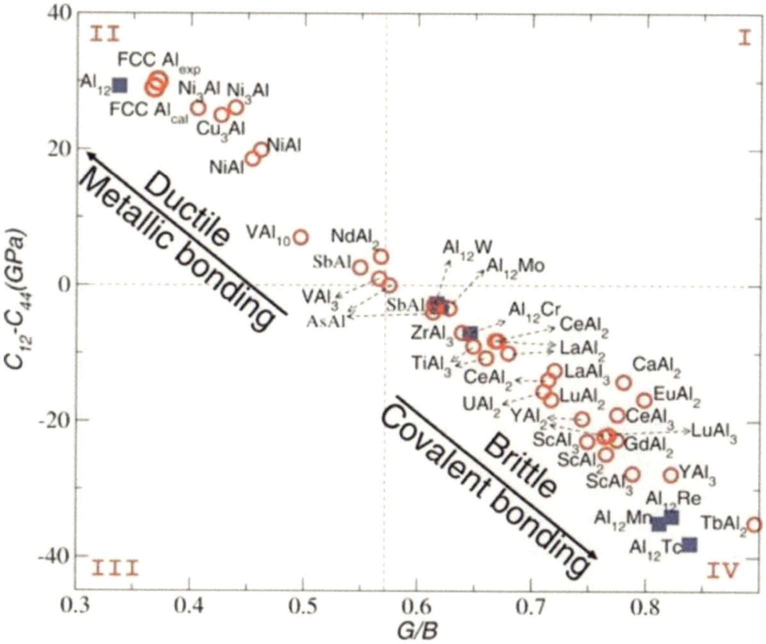

Figure 2.
Correlation between C_{12}–C_{44} and G/B for 35 aluminides (culled from [29]).

for a wide variety of aluminide group of materials is displayed in **Figure 2**. As can be seen, it is evident that an intrinsic correlation between strength and ductility of Al-based materials. It has been observed the criteria indicate a trend in a class of materials with similar deformation mechanism, but is limited by the effects of specimen sizes and crystal structures on deformation processes.

Several authors have studied elastic softening behaviour and recent evidence suggests elastic moduli manifest array of trends with a range of properties including mechanical such as hardness, yield strength, toughness and fragility [22, 30]. In early 1950s, Gilman and Cohen [31] made a historic revelations when they observed that there is a linear correlation between the hardness and elasticity in polycrystalline materials. Nevertheless, successive studies demonstrated that an uniformed linear correlation between hardness and bulk modulus does not really hold for a variety of materials [29] as illustrated in **Figure 3(a)**. Following this, Tester [32] proposed a better empirical link between hardness and shear modulus (G), as illustrated in **Figure 3(b)**. Although, the link between hardness and elastic shear modulus can be arguable, it is certain that he had demonstrated that the shear modulus, the resistance to reversible deformation under shear strain, can correctly provide a key assessment of hardness or ductility criteria for some materials. It is well known that some phase exhibits more hardness or ductility properties than others. Accordingly, it is fair to say that such descriptions could lead to further outlandish discovery in connections with regards to phase components in polycrystalline solids.

4.2 Elastic moduli and martensitic transformation

Martensitic transformation (MT) is a first-order phase type of transformation from a high-symmetry phase (austenite) at high temperature to a crystallographically low-symmetry phase (martensite) at low temperature. Martensitic behaviour has been extensively studied for decades because of its importance

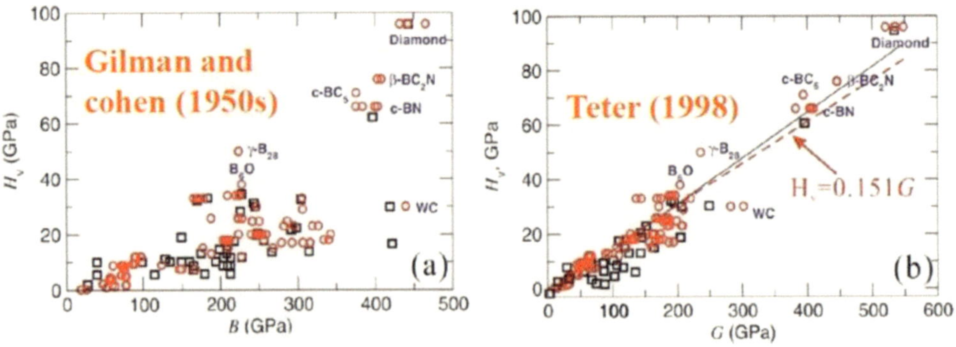

Figure 3.
Correlation of experimental Vickers hardness (H_v) with (a) bulk modulus (B) and with (b) shear modulus (G) for 39 compounds [29].

in metallurgy and its key role in shape memory phenomenon. Shape memory alloys (SMA) are materials such as TiNi and TiNi-based alloys [33], Ti-Nb [34], Ti-Mo [20, 31] etc. that exhibit diffusion-less first-order martensitic phase transitions induced by the change of temperature and/or stress. The relation between softening of elastic constants and martensitic transformation has attracted considerable attention for many years and has been discussed by many researchers [35, 36]. This interesting feature of martensitic transformation in shape memory alloys is the existence of precursor phenomena [1, 2]. The relations between MT temperature and elastic constants were investigated by Ren et al. [36]. Experiments [37] indicate that martensitic transition occurs at almost constant values of C'. Slight change in composition would cause strong deviation in the critical temperature at which C' softens to a critical value and martensitic transition occurs. In some alloys exhibiting martensitic transformation, softening of elastic constants $C' = (C_{11} - C_{12})/2$ and large elastic anisotropy, $A = (C_{44})/C'$ was observed in the parent phase, but the significance of the softening is largely different between the alloys. For example, Earlier Takashi Fukuda and co-workers [34] observed the value of C' near the transformation start temperature is approximately 0.01 GPa in In-27Ti (at %) alloy [38], 1 GPa in Au-30Cu-47Zn (at %) alloy [37], 5 GPa in Fe-30Pd (at %) alloy [39], 8 GPa in Cu-14Al-4Ni (at %) alloy [9], and 14 GPa in Ti-50.8Ni (at %) [33] and Al-63.2Ni (at %) alloys [40]. Because of such a large distribution of C' at the Ms temperature, the influence of softening of C' on martensitic transformation is expected to be significantly different between these alloys. Martensitic transformation in some alloys is probably strongly related to the softening of C', while that in others is weakly related despite the fact that the softening appears before the transformation.

Previously, Zener [5] established a correlation between the magnitudes of $C' = (C_{11} - C_{12})/2$ elastic shear modulus in metallic bcc structures with the occurrence of martensitic phase transformations suggesting links with phase stability, via the atomic interactions. He observed that the large value suggests that C' is much smaller than C_{44} and that MT temperature is dominated by C' [5]. Thus, independent elastic constants are needed to characterize the material response, such as Martensitic transformations (MTs), Shape memory etc. Martensitic transformations (MTs) are often accompanied by elastic modulus softening (acoustic phonon softening) [5]. This explains the strong

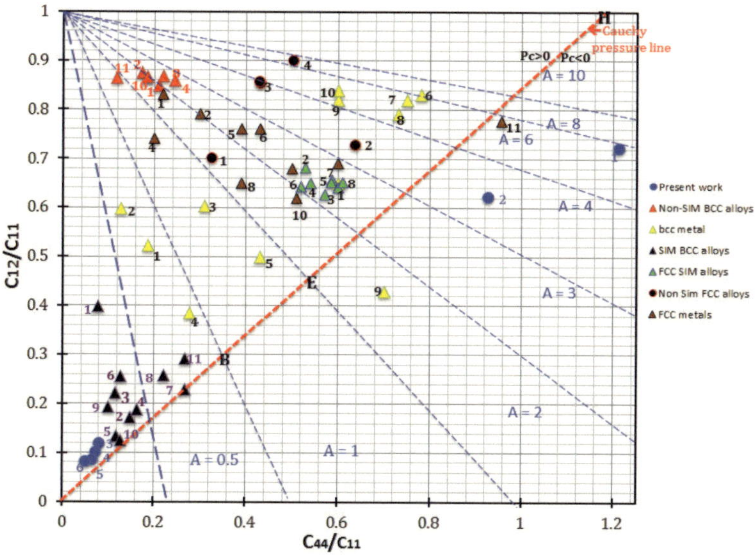

Figure 4.
Correlations between parameters reduced elastic-stiffness coefficients $(C_{12})/(C_{11})$ *vs.* $(C_{44})/(C_{11})$ *for several classes of shape memory materials (culled from [34]).*

composition-dependence of MT temperature. As a result, the modulus softens abruptly within a narrow temperature window around martensitic start temperature, Ms. However, this is unsurprising since it is well known that they are a consequence of weak restoring forces in specific crystallographic directions that announce the possibility of a dynamical instability. The elastic constants are closely related to the acoustic lattice vibrations or even atomic bondings in crystals, and accordingly will be related to the transformation mechanism for not only the Martensitic alloys but also any other compounds which accompany shear-like or displacive transitions.

Following the above, Nnamchi et al. [41] in a recent study considered the link between different groups of shape memory materials with elastic systematics found a clear delineated in a 2D plot of two dimensionless ratios of elastic constants or reduced elastic-stiffness coefficients, $(C_{12})/(C_{11})$ vs. $(C_{44})/(C_{11})$ formally popularised earlier by Blackman [42], It is only one table with different sections. (see **Figure 4** and **Tables 1** and **2**). This reveals among others the elastic anisotropy, proximity to Born mechanical instability, elastic-constants (interatomic-bonding) changes caused by alloying, pressure, temperature, phase transformations and similarities in types of interatomic bonding. The significance of the softening is largely different between the alloys. Inspecting the diagram, we notice materials with similar chemical bonding tend to fall in the same region of the diagrams. Such diagrams provide many uses.

	This work	C_{11}	C_{12}	C_{12}/C_{11}	C_{44}/C_{11}	Ref.
1	Ti-3Mo	159.3	115	0.72	1.21	[34]
2	Ti-6Mo	111.3	69.07	0.62	0.93	[34]
3	Ti-10Mo	167	19.6	0.12	0.081	[34]
4	Ti-14Mo	179.2	17.9	0.10	0.074	[34]

	This work	C_{11}	C_{12}	C_{12}/C_{11}	C_{44}/C_{11}	Ref.
5	Ti-18Mo	192.6	16.3	0.085	0.066	[34]
6	Ti-23Mo	197.5	16	0.081	0.051	[34]
	Non-SIM (BCC)alloys					
1	Ti50Ni30Cu20	209	183	0.88	0.17	[43]
2	Ti-50Ni	165	140	0.85	0.21	[44]
3	Ti-29Nb-13Ta-4.6Zr	67.1	39.9	0.87	0.19	[45]
4	Ti-30Nb-10Ta-5Zr	128	92	0.86	0.24	[46]
5	Ti-35Nb	163.5	142	0.87	0.22	[47]
6	Ti-30Nb-5Ta-5Zr	70	30	0.87	0.185	[48]
7	Ti-32.7Nb-11.6Ta-4.49Zr-0.066O-0.052N	137	91.1	0.86	0.12	[49]
	Non-SIM (FCC) alloys					
1	Ag-75Au	230	161.5	0.702	0.33	[43]
2	Cu-4.17Si	117	85.2	0.73	0.64	[43]
3	α-Ag-2.4Zn	190	162	0.85	0.43	[43]
4	α-Cu-9.98Al	199	179	0.89	0.50	[43]
5	α-Cu-22.7Zn	158.9	136.2	0.86	0.43	[43]
	SIM (BCC) alloys					
1	Ti-35.37Nb	130.2	52	0.40	0.078	[50]
2	Ti-35Nb-2Zr-0.7Ta	183	31.4	0.17	0.15	[49]
3	Ti-35.4Nb-1.9Ta-2.8Zr-0.37O	122	27	0.22	0.11	[49]
4	Ti-24.1Nb-4Zr-8.06Sn-0.15O	140	26.3	0.19	0.16	[49]
5	Ti-35Nb-10Ta-4.6Zr-0.16O	102.5	36	0.16	0.12	[51]
6	Ti-23.9Nb-3.75Zr-8.01Sn-0.04O	157.2	36	0.26	0.127	[51]
7	Ti-24Nb-4Zr-7.9Sn-0.17O			0.23	0.22	[51]
8	Ti-24Nb-4Zr-7.6Sn-0.07O	122	31.4	0.26	0.21	[49]
9	Ti-35.2Nb-10.5Ta-4.97Zr-0.091O-0.014N	140	27	0.19	0.1	[51]
10	Ti-23.9Nb-3.8Zr-7.61Sn-0.08O	102.5	26.3	0.12	0.13	[51]
11	Ti-24Nb-4Zr-7.9Sn	157.2	46	0.29	0.27	[52]
	SIM (FCC) alloys					
1	Cu44.9- 50Zn	125	80	0.64	0.6	[43]
2	Au47.5-50Cd	142	96.77	0.68	0.53	[43]
3	Ag45-50Zn	132.8	83.16	0.63	0.57	[43]
4	γ-FeNi	209	183	0.65	0.54	[53]
5	CuAlNi	142.8	93.7	0.66	0.59	[54]
6	B2-NiTi	162	104	0.64	0.52	[55]
7	Cu2.726A11.122Ni 0.152	137	89.2	0.65	0.59	[56]
8	Cu2.742Al1.105Ni0.152	136	81.763	0.65	0.61	[56]

Table 1.
Elastic constant of some bcc and fcc metals and alloys.

	BCC elements	C12/C11	C44/C114	Ref.
1	V	0.52	0.19	[53]
2	Nb	0.59	0.13	[53]
3	Ta	0.60	0.31	[53]
4	Mo	0.38	0.28	[53]
5	W	0.5	0.43	[53]
6	Li	0.83	0.78	[53]
7	Na	0.82	0.75	[53]
8	K	0.79	0.73	[53]
9	Ba	0.43	0.7	[53]
	FCC elements			[53]
1	Au	0.83	0.22	[53]
2	Pd	0.79	0.3	[53]
3	Pt	0.74	0.2	[53]
4	Ag	0.76	0.39	[53]
5	Cu	0.76	0.43	[53]
6	β-Co	0.69	0.6	[53]
7	α-Sr	0.65	0.39	[53]
8	γ-Fe	0.68	0.5	[53]
9	Ni	0.62	0.51	[53]
10	δ-Pu	0.78	0.96	[53]

Table 2.
Elastic constant of some bcc and fcc metals and alloys.

5. Summary and future challenges

The following bullet points summarise some of the main challenges facing the community.

- Some empirical elastic relationship such as a low G/B ratio (or high ν) favours toughness but also indicates a fragility in polycrystalline materials, though they can be typically difficult to vitrify in some polycrystalline materials.

- Some empirical correlations exist in most of the metallic elements in the periodic table have been found, and alloy development has moved beyond the bucket chemistry type approach used in the early days of elastic properties research. While a number of general guidelines exist for explaining elastic systematics property formation (such as Zener, and Burger's rules), Pugh and Pettifor's criterion [16, 17] in addition to Blackmans have gone beyond simply stating the chemical species that should be present, and their rough proportions, and instead gives exact elastic relationship. However, a more rigorous that delineated the phase stability using systematics could be envisaged in new future.

Glossary of symbols

Symbols used symbols derived from disambiguation (e.g. d for

C_{ijkl}	the local elastic constant tensor with $\langle\sigma_{ij}(\boldsymbol{r})\rangle$ and $\langle\epsilon_{kl}(\boldsymbol{r})\rangle$ as the local stress and strain field at a point **r**, respectively, and the angular brackets denote ensemble averages
C_{44}	single crystal bulk modulus; $B^o = (C_{11}^0 + 2C_{12}^O)/3$
$C' = (C_{11-}\ C_{12})/2$	tetragonal shear modulus
$C_{44}, \mu^* = C_{44}^* = \mu^0 + \frac{\langle\tau_{44}\rangle}{1+G_{44}\tau_{44}}$	trigonal shear modulus
G	the ratio of shearing stress τ to shearing strain γ within the proportional limit of a material
B	bulk modulus, ratio between the fluid pressure and the Volumetric Strain
E	modulus of elasticity or Young's modulus
G	modulus of rigidity or shear modulus
V_L and V_S	the ultrasonic longitudinal and shear wave velocities respectively
ρ	the density of the material
$A = (C_{44})/C^{'}$	elastic anisotropy
U	the energy of the crystal, and quadratic function of the strains
V_0	equilibrium volume
e	an elastic strain
$\langle\sigma_{ij}(\boldsymbol{r})\rangle$	effective stiffness of C_{ijkl}
$\epsilon = \epsilon^0 + GT\epsilon^0$	ϵ^0 and GT are the strain and modify Green's function
T	T-matrix is given by $T = \delta C(1 - G\delta C)^{-1}$
$\boldsymbol{I}$	equivalent to the unit tensor
$\widetilde{Y}*$	the homogenised polycrystalline Young's modulus
$\widetilde{\mu}*$	homogenised polycrystalline Poisson's ratio
M_S	martensite formation start temperature
M_F	martensite finish temperature
SME	shape memory effect

Author details

Paul S. Nnamchi[1,2*] and Camillus S. Obayi[2]

1 Department of Mechanical and Construction Engineering, Northumbria University, Newcastle, United Kingdom

2 Department of Metallurgical and Materials Engineering, University of Nigeria, Nsukka, Enugu State, Nigeria

*Address all correspondence to: nnamchi.paul@gmail.com

References

[1] A History of the Theory of Elasticity and of the Strength of Materials Vol. 1 (1886) Vol. 2 (1893)

[2] Born M. Mathematical Proceedings of the Cambridge Philosophical Society. 1940;**36**:160

[3] Born M, Huang K. Dynamics Theory of Crystal Lattices. Amen House, London: Oxford University Press; 1954

[4] Nye JF. Physical Properties of Crystals. Oxford, UK: Oxford University Press; 1985

[5] Zener JC. Elastic and Anelastic of Metals. Chicago, Illinois, USA: University of Chicago press; 1948

[6] Pagnotta L. Recent progress in identification methods for the elastic characterization of materials. International Journal of Mechanics. 2008;**2**(4):129-140

[7] Pokluda J et al. Ab initio calculations of mechanical properties: Methods and applications. Progress in Materials Science. 2015;**73**:127-158

[8] Nye JF. Physical Properties of Crystals: Their Representation by Tensors and Matrices. Amen House, London: Oxford University Press; 1985

[9] Grimvall G, Magyari-Kope B, Ozolin V, Persson KA. Lattice instabilities in metallic elements. Reviews of Modern Physics. 2012;**84**:945-986

[10] Mouhat F, Coudert F-X. Necessary and sufficient elastic stability conditions in various crystal systems. Physiological Reviews. 2014;**B90**:224104

[11] Milman V, Warren MC. Journal of Physics: Condensed Matter. 2001;**13**:241

[12] Rajamallu K, Niranjan MK, Dey SR. Materials Science and Engineering: C. 2015;**50**:52-58

[13] Zeller R, Dederichs PH. Physica Status Solidi B. 1973;**55**:831-842

[14] Middya TR, Basu AN. Journal of Applied Physics. 1986;**59**:2368-2375

[15] Middya TR, Paul M, Basu AN. Journal of Applied Physics. 1986;**59**: 2376-2381

[16] Raabe D, Sander B, Friak M, Ma D, Neugebauer J. Acta Materialia. 2007;**55**: 4475-4487

[17] Eriksson O. Encyclopaedia of Materials: Science and Technology. Amsterdam: Elsevier; 2003. pp. 1-11

[18] Beznosikov BV. Journal of Structural Chemistry. 2003;**44**:885

[19] Thompson JR, Rice JR. Philosophical Magazine Letters. 1974;**29**:269-277

[20] Nnamchi PS. Materials and Design. 2016;**108**:60-67

[21] Wang WH. Journal of Non-Crystalline Solids. 2005;**351**:1481-1485

[22] Lewandowski JJ, Wang WH, Greer AL. Philosophical Magazine Letters. 2005;**85**:77-87

[23] Pugh SF. XCII. Relations between the elastic moduli and the plastic properties of polycrystalline pure metals. The London, Edinburgh, and Dublin Philosophical Magazine and Journal of Science. 1954;**45**:823-843

[24] Pettifor DG. Theoretical predictions of structure and related properties of intermetallics. Materials Science and Technology. 1992;**8**:345-349

[25] Pugh SF. Philosophical Magazine. 1954;**45**:834

[26] Thompson RP, Clegg WP. Predicting whether a material is ductile

or brittle. Current Opinion in Solid State & Materials Science. 2018;**22**(3):100-108. DOI: 10.1016/j.cossms.2018.04.001

[27] Cottrell AH. In: Charles JA, Smith GC, editors. Advances in Physical Metallurgy. London: Institute of Metals; 1990. pp. 181-187

[28] Poon SJ, Zhu A, Shiflet GJ. Applied Physics Letters. 2008;**92**:261902

[29] Tester DM. MRS Bulletin. 1998; **23**:22

[30] Gilman JJ, Cohen. Science. 1993; **261**:1436-1439

[31] Gao FM, Gao LH. Journal of Superhard Materials. 2010;**32**:148

[32] Ren X. Materials Science and Engineering. 2001;**A312**:196-206

[33] Fukuda T, Kakeshita T. Metals. 2017;7:156. DOI: 10.3390/met7050156

[34] Nnamchi PS, Todd I, Rainforth MW. Novel approach to Ti alloy design for biomedical applications. UK: Department of Materials Science and Engineering, University of Sheffield; 2014

[35] Murakami Y. Lattice softening, phase stability and elastic anomaly of the β-Au-Cu-Zn alloys. Journal of the Physical Society of Japan. 1972;**33**: 1350-1360

[36] Gunton DJ, Sanuders GA. The elastic behavior of In-Tl alloys in the vicinity of the martensitic transformation. Solid State Communications. 1974;**14**:865-868

[37] Michal L et al. Materials Science and Engineering A. 2007;**462**:320-324

[38] Scripta Materialia. 1998;**38**(11): 1669-1675

[39] Muto S, Oshima R, Fujita FE. Elastic softening and elastic strain energy consideration in the fcc-fct transformation of FePd alloys. Acta Metallurgica et Materialia. 1990;**38**: 685-694

[40] Enami K, Hasunuma J, Nagasawa A, Nenno S. Elastic softening and electron-diffraction anomalies prior to the martensitic transformation in Ni-Al β1 alloy. Scripta Metallurgica. 1976;**01**: 879-884

[41] Nakanishi N. Elastic constants as they relate to lattice properties and martensite formation. Progress in Materials Science. 1980;**24**:143-265

[42] Blackman M. Proceedings. Royal Society of London. 1938;**164**:62

[43] Ren X et al. Philosophical Magazine A. 1999;**79**:31-41

[44] Hatcher N, Kontsevo Yu O, Freeman A. Physical Review B. 2009;**80**: 1-18

[45] Li SJ, Cui TC, Hao YL, Yang R. Acta Biomaterialia. 2008;**4**:305-317

[46] Nobuhito S, Niinomi M, Akahori T. Materials Transactions. 2004;**45**(4): 1113-1119

[47] Zhengjie L, Wan L, Xiaobing X, Weijie L, Qina J, Zhang D. Materials Science and Engineering: C. 2013;**33**: 4551-4561

[48] Hou FQ, Li SJ, Hao YL, Yang R. Scripta Materialia. 2010;**63**:54-57

[49] Tane M, Nakano T, Kuramoto S, Hara M, Niinomi M, Takesue N, et al. Acta Materialia. 2011;**59**:6975-6988

[50] Kim HY, Ikehara Y, Kim JI, Hosoda H, Miyaaki S. Acta Metallurgica. 2006; **54**:2419

[51] Talling RJ, Dashwood RJ, Jackson M, Kuramoto S, Dye D. Scripta Materialia. 2008;**59**:669

[52] Hao Y, Li S, Sun B, Sui M, Yang R. Physical Review Letters. 2008;**98**:1-4

[53] Paszkiewicz, Wolski. Journal de Physique(Conference series). 2008;**104**: 012038

[54] Mañosa L et al. Physical Review B. 1994;**49**:9969-9972

[55] Zhou L, Cornely P, Trivisonno J, Lahrman D. An ultrasonic study of the martensitic phase transformation in NiAl alloys. Honolulu, HI, USA: IEEE Symposium on Ultrasonics; 1990; **3**:1389-1329

[56] Sedlak P, Seiner H, Landa M, Novak V, Sittner P, Li M. Acta Materialia. 2005; **53**:3643-3661

Chapter 5

Repair Inspection Technique Based on Elastic-Wave Tomography Applied for Deteriorated Concrete Structures

Katsufumi Hashimoto, Tomoki Shiotani, Takahiro Nishida and Nobuhiro Okude

Abstract

Applying elastic wave tomography as an innovative NDT method, the evaluation of velocity distribution in three-dimensional (3D) before and after the repair is introduced in this study. The increase in the velocity with penetration of the repair material according to the repair effect is identified visually and quantitatively. The 3D tomography technique is newly proposed for one-side access inspection, using drill hammering to generate an elastic wave. Accordingly, the elastic wave velocity distribution result enables to visualize the internal quality of concrete after patch repair is successfully done. In addition, an attempt for reinforced concrete (RC) slab panels is made to confirm the effectiveness of the repair by comparing the velocity distribution of elastic waves obtained from acoustic emission (AE) tomography analysis, before and after the repair. Thus, the velocity recoveries due to injection are found in all the slab panels, and it is confirmed that the elastic wave velocities obtained using this technique can serve as an indicator for examining the state of crack and void filling with injected material. Further, a good correlation is found between the low-velocity region before repair and the amount of injection. These results show the potential of the AE tomography technique to be used as a method for estimating the effect of injection repair.

Keywords: elastic wave, acoustic emission, wave velocity distribution, tomography, repair method

1. Introduction

It is highly demanded to establish sufficient management systems for the inspection of existing concrete infrastructures in order to manage and extend their service lives. As for aging infrastructure, severe deterioration is currently reported, where it is known as a critical issue in our society, and large budgets are required to repair damaged structures. Since budgetary restrictions are often imposed, preventive and proactive maintenance techniques of infrastructure are sufficiently needed with nondestructive testing (NDT) methods. In addition to conventional NDT, innovative methods must be established to appropriately assess and evaluate damage and repair and retrofit recovery in concrete structures. Inspection techniques after crack repair methods application for existing structures to assess repair installations have not yet been practically developed, meanwhile improper repair efforts have resulted in re-deterioration. Refilling internal cracks with repair materials from the concrete surface, epoxy injection, and patch repair methods are widely implemented. In most cases, re-deterioration could be led by the unknown and remained internal defects. Consequently, it is very important to implement and establish inspection techniques which can visualize internal defects as a countermeasure with repair works.

For such infrastructure as bridges and tunnels, it is generally recognized that appropriate maintenance works are necessary. Prior to extensive damage and failure in existing structures, essential issues include establishing a maintenance system for reinforced concrete (RC) members with the sufficient measures. Epoxy injection and patch repair methods have been widely and practically introduced to repair and re-strengthen RC members. However, insufficient repair works are unfortunately often reported, and these works have potentially resulted in re-deterioration because more improvement is needed for inspection techniques to estimate the quality of repair and recovery.

Developing nondestructive testing and evaluation methods is strongly demanded for concrete structures to quantify or assure the repair and retrofit recovery. The International Union of Laboratories and Experts in Construction Materials, Systems, and Structures (RILEM) launched a technical committee on innovative NDT for repair and retrofit recovery [1]. Tomography techniques are studied based on elastic wave and acoustic emission (AE) to visualize, internal defects in three-dimension concrete with the committee's activities. These techniques applicability has already been published in terms of elastic wave tomography [2, 3] and AE tomography [4, 5].

Using parameters of elastic wave such as amplitudes and elastic wave velocities, internal distributions are obtained by the tomography technique. Elastic wave velocity is specifically used as the parameter in this study. Both the location of the excitation and the excitation time are known in the mentioned elastic wave tomography. On the other hand, they are unknown for AE tomography. The elastic wave velocity in each set-element over the structure can be calculated. Elastic wave velocity is theoretically associated with elastic modulus of material. The values would vary as low-velocity zones with the presence of such internal defects as cracks and voids.

In a theory of elastic wave propagation inside media, the waves are reflected, diffracted, and scattered where it has voids and cracks. Elastic wave velocity is known to be decreased by the phase divergence. The zones of lower elastic wave velocity corresponding to those of heavier deterioration can be reasonable assumed. The distribution of wave velocities can be accordingly referred to as a good indicator of the internal condition of a concrete structure. Moreover, in order to guarantee whether the injected material is properly filled into cracks by using the crack injection method, the velocity distributions of elastic waves in the applicable regions of RC structures are estimated, before and after the repair, by employing AE tomography method [6].

The repair effects in concrete were evaluated with 3D elastic wave tomography in the present study by means of innovative NDT, which can visually identify the outcome from the repair condition provided by the epoxy injection and patch repair methods. 3D tomography was employed for a 50-year-old concrete pier, which was repaired by epoxy injection method, as well as to a 53-year-old concrete wall, which was repaired by the patch repair method. And, AE tomography was applied to a 46-year-old RC slabs, in which epoxy-based resin was used as the injected material to repair the internal cracks.

As described here, although the epoxy injection and patch repair methods are major repair methods even without the corrosion of the reinforcing bars, there are many reports indicating re-deterioration with insufficient repairs. This study aims to validate the 3D elastic wave tomography and AE tomography technique for inspection of the internal quality of concrete after repair.

2. Examples of existing concrete structures for repair inspection

2.1 Pier

Concrete pier specimen, 600 mm width, 1200 mm height, and 300 mm thickness, is shown in **Figure 1**. About 93 components of syringe-type caulking guns were set into pots for injection and 50 kHz resonance AE sensors were arrayed to receive elastic waves before the injection and 7 days after injection, which is corresponding to the epoxy resin hardening period.

Attached AE sensors to four sides of the pier, as shown in **Figures 2** and **12** sensors were arranged on sides A and B in a 600 × 1200 mm area and 4 sensors were installed on the other sides. About 25 mm diameter steel ball was used for the excitation of elastic wave. In order to identify the impact excitation time, at the closest sensor location, each excitation point was selected.

2.2 Wall

Figure 3 shows concrete wall, 600 × 600 mm, where the patch repair method was applied, following V-shaped concrete removal was conducted for 80 mm depth and 120 mm width since surface cracks with water leakage were observed on the

Figure 1.
Overview of concrete pier.

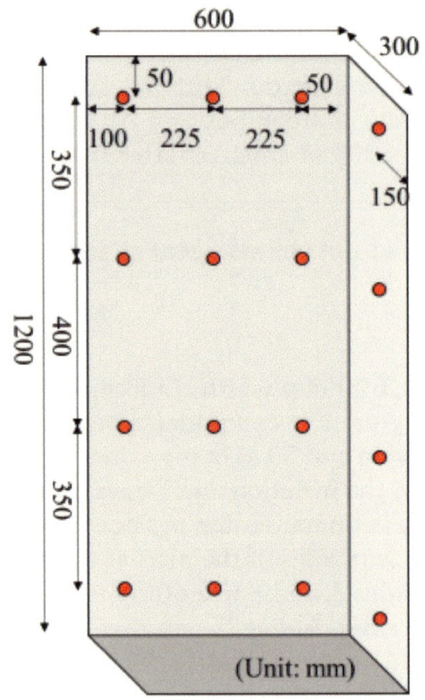

Figure 2.
Sensor arrangement.

Figure 3.
Overview of concrete wall.

surface of the tunnel-lining concrete. Then, polymer cement mortar with a water-to-cement ratio W/C = 25% was used to fill the crack. Employing micro-core drilling and hammering as one-sided access measurement, the wave signals generated inside the concrete were detected. A 12 mm diameter micro-coring was performed up to 200 mm depth. A curved edge 6 mm diameter steel bar was inserted into the bit hole. The head of steel bar was hit by 25 mm diameter spherical steel ball. Hammering the steel bar without touching the hole wall, elastic waves could only be generated at the hole end in the depth direction. About 60 kHz resonance AE sensors were installed to detect the elastic waves. The sensor arrangements and excitation points are shown in **Figure 4**.

2.3 Slab

Figure 5 and **Table 1** show a top view of an RC bridge, and specifications for the measured deck panels. This bridge is a municipal road bridge located in the Hokuriku region, Japan and it has been in service in the last 46 years. Three panels

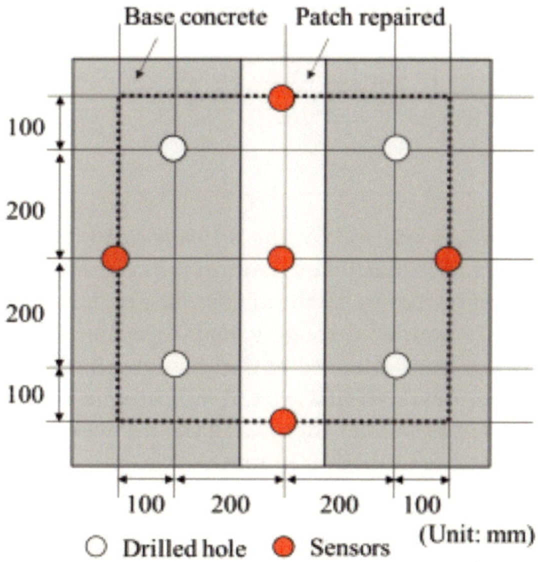

Figure 4.
Locations of drilling and sensor arrangement.

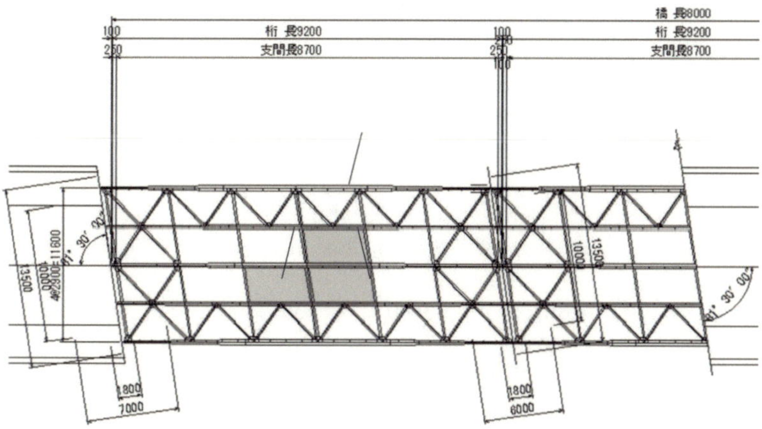

Figure 5.
A top side view of subject bridge.

Type	**RC bridge (3 span composite girder bridge)**
Length	88.0 m
Age	46 years
Thickness	Slab: 250 mm and asphalt: 50 mm
Condition	Web-shaped cracks were sporadically evident on the concrete surface.

Table 1.
Bridge specifications.

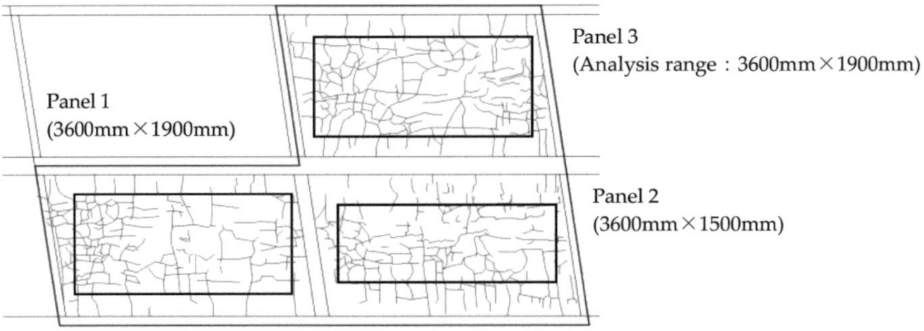

Figure 6.
Sketch of cracking.

highlighted in the figure are selected for the measurement. On all of the slab panels, web-shaped cracks were sporadically evident on the concrete surface. These cracks are thought to be caused primarily by the alkali-silica reaction in concrete. **Figure 6** shows a sketch of cracks obtained through visual inspection from the bottom side of the slab. This figure also shows the area of the tomography analysis for obtaining the velocity distribution. Crack widths are not indicated in figure, but in all the slab panels, the cracks width was smaller than 0.2 mm, and over almost the entire range, the widths were in the range of 0.10–0.15 mm.

3. Data analysis

3.1 Wave detection and computation of elastic wave arrival time

In order to determine the velocity distributions by tomography, the following analytical steps are taken.

First, the arrival time at each sensor was determined with an Akaike Information Criterion (AIC) picker [7, 8]. For the digitized wave record x_k of length N, the AIC value is defined as

$$\begin{aligned} \mathrm{AIC}(k) = k \times \log\{\mathrm{var}(x[1,k])\} \\ + (N-k-1)*\log\{\mathrm{var}(x[k+1,N])\} \end{aligned} \tag{1}$$

where var$(x[1, k])$ indicates the variance between x_1 and x_k, and var$(x[k, N])$ is the variance between x_k and x_N.

The point where AIC value minimizes, applying the least-square method, corresponds to the most suitable separation point of two series of stationary time, the arrival time as the phase onset is thus reasonably determined by the AIC picker. Lower AIC values suggest noise and higher AIC values show the arrival of wave signals. Following the determination of arrival time, the elastic wave velocity is calculated. The observed time of wave propagation T_{obs} is obtained by [9].

$$T_{obs} = T_o - T_s \tag{2}$$

where T_s is the time of excitation and T_o is the arrival time.

3.2 Elastic wave tomography

The reciprocal of the velocity is referred in the elastic wave tomography algorithm to as the "slowness." As shown in **Figure 7**, slowness as the initial parameter

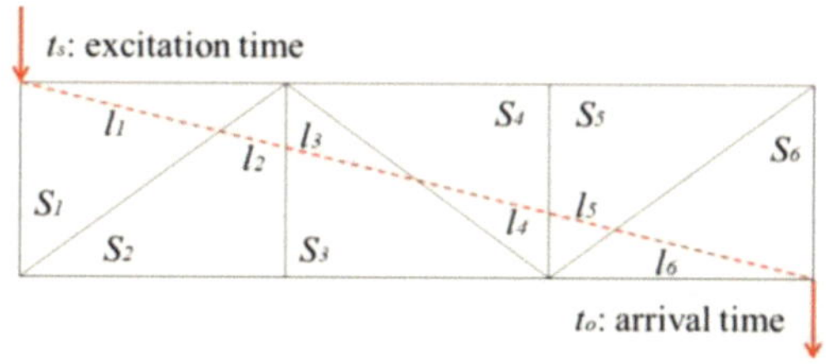

Figure 7.
Slowness for calculation of propagation time.

is provided into each element. Travel time of elastic wave can be computed as elastic velocity is constant in individual element on this ray path. The total of the propagation time calculated by the slowness and the distance in each element (refer to Eq. (3)) derives the propagation time T_{cal}. The difference between the observed propagation time (T_{obs}) and the theoretical propagation time (T_{cal}) is obtained by Eq. (4).

where l_j is the length crossing each element and s_j is the slowness of each element.

$$T_{cal} = \sum_j sj*lj \tag{3}$$

$$\Delta T = T_{obs} - T_{cal} \tag{4}$$

s_i is slowness of element i, l_i is length of the ray path in element i. Thus, it is revealed that l_i is essential for the calculation of the travel time.

In order to reduce the difference between the observed propagation time and the theoretical propagation time, the slowness in each element is re-calculated and renewed. The total slowness correction is determined by Eq. (5) and the revised slowness is consequently calculated by Eq. (6).

$$\begin{bmatrix} \Delta s1 \\ \Delta s2 \\ \vdots \\ \Delta sj \end{bmatrix} = \begin{bmatrix} \sum_i \frac{\Delta T_i{}^* l_{i1}}{L_i} \Big/ \sum_i l_{i1} \\ \sum_i \frac{\Delta T_i{}^* l_{i2}}{L_i} \Big/ \sum_i l_{i2} \\ \vdots \\ \sum_i \frac{\Delta T_i{}^* l_{ij}}{L_i} \Big/ \sum_i l_{ij} \end{bmatrix} \tag{5}$$

$$s'j = sj + \Delta sj \tag{6}$$

where L_i is the total distance of wave propagation through the i-element.

Proceeding the iteration based on Eqs. (5) and (6) as shown in **Figure 8**, the optimal slowness, eventually the velocity, in each element corresponding to the observed propagation times of multiple paths over the interested area is determined as well as the velocity distribution.

In order to determine the ray path more accurately, the ray trace algorithm is applied, taking into account detours of elastic waves due to the reflection and diffraction. Following 3D ray trace algorithm, which was proposed in previous research [3], the arrival time of each wave is obtained. Correction of the slowness in each element is carried out according to the error between the observed first travel time and computed value in the element, using 3D finite elements for meshing of target space in the present algorithm. Wave velocities between 2000 and 4500 m/s are given for the tomography results as the range of wave velocities in concrete.

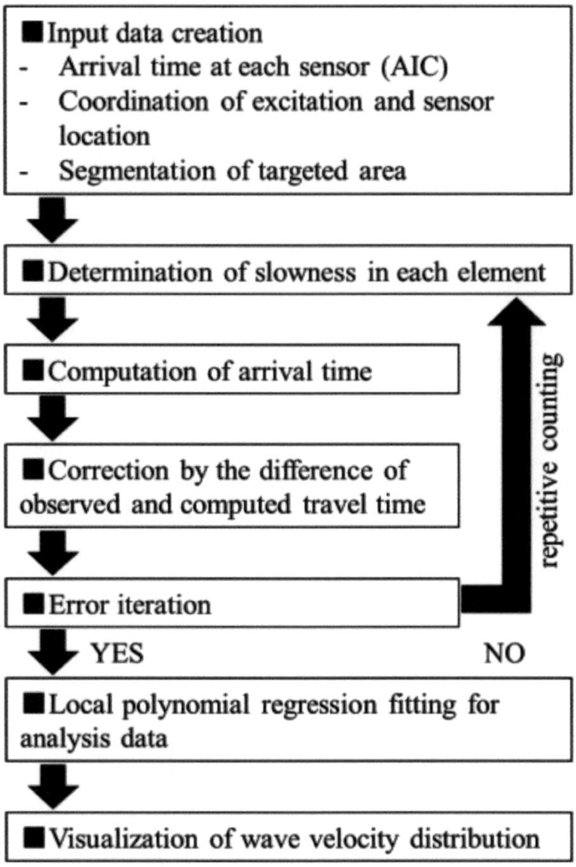

Figure 8.
Analytical procedure for 3D tomography.

3.3 Acoustic emission (AE) tomography

AE tomography is a method for obtaining a velocity distribution by finding the travel time from an AE source to each sensor. Thus, it is necessary to obtain the position of the transmission source as accurately as possible. With the conventional ranging technique, which assumes that the propagation velocity is fixed, considerable errors are expected in the case that the tomography technique is applied to such a heterogeneous material as concrete. Consequently, a new ranging technique incorporating with the ray tracing concept has been developed as a pre-processing technique for AE tomography [6]. The ranging technique using ray tracing is illustrated in **Figure 9**. As shown in the diagram, ray tracing is performed from the received point j to all other nodes i, and the theoretical travel time T_{ji} to each node is calculated. The shortest transmission time is determined from the differences between T_{ji} and the initial travel time T_j at the received point j. The procedure is repeated for the number of received points N, and finally the node, where the variance of estimated arrival times estimated from Eqs. (7) and (8) becomes the minimal, is taken to be the transmission point. In Eq. (7), T_{mi} is the mean value of the estimated transmission times at each node i, and in Eq. (8), σ_i is the variance of the estimated transmission times at each node i.

$$T_{mi} = \frac{\sum_j (T_{ji} - T_j)}{N} \tag{7}$$

$$\sigma_i = \frac{\sum_j \left(T_{ji} - T_j - T_{mi}\right)^2}{N} \tag{8}$$

3.4 Excitation method of elastic waves and analysis model

Figure 10 shows the model of AE tomography analysis and the positions of receiving sensors. The shaded part at the top of the model indicates the asphalt layer (thickness: 50 mm). Analyzed regions for slab panels 1 and 3 were set to be 3600 × 1900 mm. Concerning slab panel 2, there were limitations on the sensor positions, and thus the region was set to be 3600 × 1500 mm. As elements for AE tomography analysis, the applicable region was divided by 16 × 8 in total of 128 elements. In AE tomography, elastic waves were excited by the steel ball drop. A steel ball of 5 mm diameter was dropped at several locations for 12 minutes from the asphalt surface, consciously ensuring that the distribution of impact points was as uniform as possible at the target area. The steel ball dropping is illustrated in

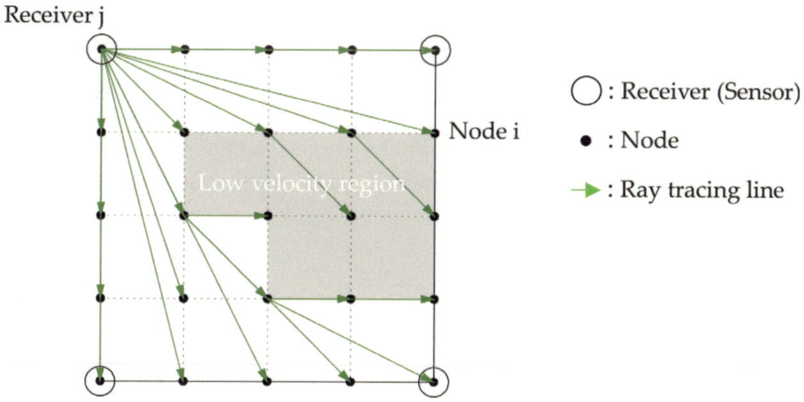

Figure 9.
Overview of transmission source estimation using ray tracing.

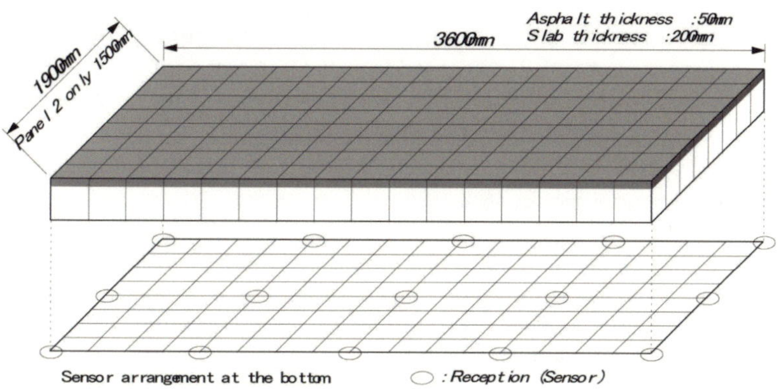

Figure 10.
Analysis model for AE tomography.

Figure 11.
Steel ball dropping.

Figure 11. In AE tomography, the measurements were performed using an acceleration measurement system (TEAC). About 15 piezoelectric accelerometers with the frequency response from 3 Hz to 15 kHz were employed as receiving sensors. The point at which AIC is the minimum is determined as the arrival time of the wave. However, when the S/N ratio is low, it is difficult to identify the minimum value of AIC. Thus, a reliability parameter is developed for reading the initial travel time. The index is proposed as a measure for the identification of the rising edge of the wave [10]. It is found that readings of the initial travel times reasonably converge if the index is 0.05 or higher. In the present chapter, elastic waves with the index of 0.1 or higher are analyzed. AIC (k_{min}) indicates the minimum value of AIC, that is, corresponding to the initial travel time.

4. Results and discussion

4.1 Pire

4.1.1 Wave detection and computation of elastic wave arrival time

In order to investigate the epoxy-injected situation in damaged concrete, black light (ultraviolet light) was irradiated on the cored sample so that the injected material (epoxy resin) was colored in blue as shown in **Figure 12**. It is confirmed that epoxy resin was successfully injected into the concrete cover (up to 10 cm) and over the depth of the reinforcing bars (from 10 to 15 cm). The injected material can penetrate cracks even smaller than 0.1 mm in width [11].

4.1.2 3D elastic wave tomography

Figure 12 shows elastic wave tomography results before and after epoxy injection. The overview of injection repair method is shown in **Figure 13**. The wave velocities after the repair indicate clearly higher values than those before. The velocities are mostly higher than approximately 3000 m/s. This result implies that the epoxy injected from the surface of the pier could be filled and hardened sufficiently inside the media via cracks. However, at the central portions of the concrete pier, wave velocities are still lower than 2500 m/s, namely, the epoxy injection only guarantees the shallow zone repair from the concrete surface. The velocity

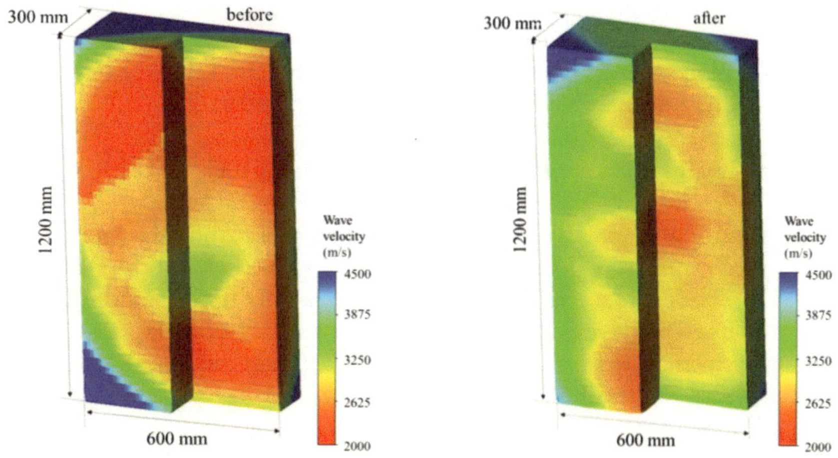

Figure 12.
Results of elastic wave tomography in 3D (left: before injection and right: after injection).

Figure 13.
Repair by epoxy injection.

distribution given by 3D elastic wave tomography shows the conditions inside the concrete, in particular, whether the epoxy is fully penetrated into the interior, while it is noted that the tomography technique could assess the repair level, which is not visually clarified on the exterior.

Figure 14 shows wave velocity histogram before and after the repair consequences. The mean value after the repair is higher than that before, and the variation decreases. Since the velocities lower than 2500 m/s are rarely observed in the histogram, concrete of the pier is repaired after the injection.

4.1.3 Amount of epoxy injection and alteration of wave velocities

Figures 15 and **16** respectively show the tomography results the injected epoxy amount at Side A (referred as the front surface in **Figure 12**) and those at Side B (referred as the back surface in **Figure 12**). They are only the tomograms of wave velocities at the surface layer, comparing with the amount of epoxy injection. The injection pots which added syringe refilling are colored in red because the caulking guns were replaced and refilled with epoxy until the spring-loaded gun automatically stopped the injection.

The velocity distribution alteration reasonably correlates with the epoxy injection amount. Concrete property improvement suggested by the velocity recovery is also roughly confirmed with the epoxy injection amount.

Velocity-improved areas in **Figures 15** and **16** are relatively observed in the areas, where additional injection was installed because of their porous media due to heavy deterioration (red colored in **Figures 15(d)** and **16(d)**). Less improvement of the velocity are observed even after the repair at the bottom-right corner of side A (see **Figure 15**), where the method did not enable to penetrate the resin sufficiently into the concrete because of their lower connectivity of internal cracks.

4.2 Wall

3D elastic wave tomography technique mentioned above was challengingly applied to confirm patch repair effect for concrete wall of an existing structure. In this study, the technique was introduced as a method to evaluate the retrofit recovery. There is currently no NDT technique applicable in terms of in-situ measurement.

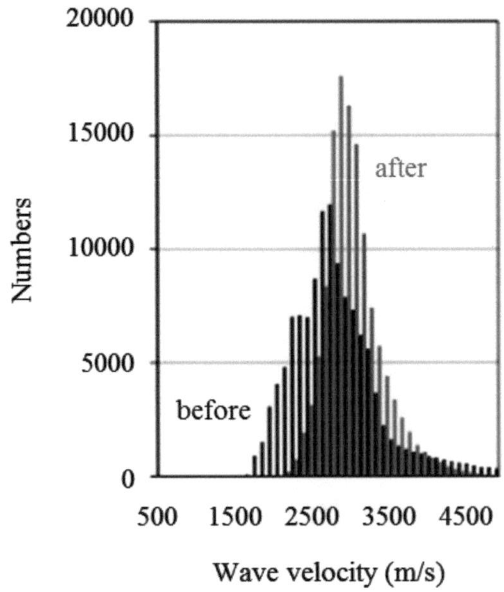

Figure 14.
Histogram of wave velocities.

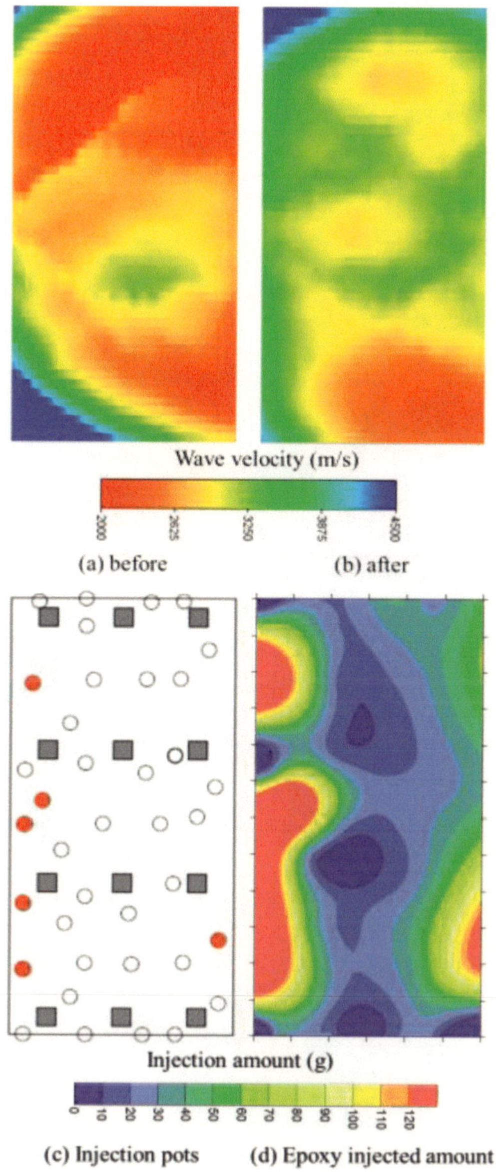

Figure 15.
Results of wave velocities and amount of epoxy injection (Side A).

4.2.1 Procedures of drilling and excitation

Introducing micro-core drilling, excitation of elastic waves was driven. The technique is proposed and applied usefully for one-side access inspection works. **Figure 17** shows the illustration of test procedure schematically.

On the surface of concrete wall with a surface crack (a), a V-cut concrete removal is performed (b), followed by a patch repair method with polymer cement mortar grouting (c), 12 mm diameter bit hole of 200 mm depth is drilled by micro-coring (d), at each concrete surface point. With the sensor array on the surface (e), 6 mm diameter a steel bar is inserted into the hole and the steel bar head is hit by a 25 mm diameter steel sphere ball.

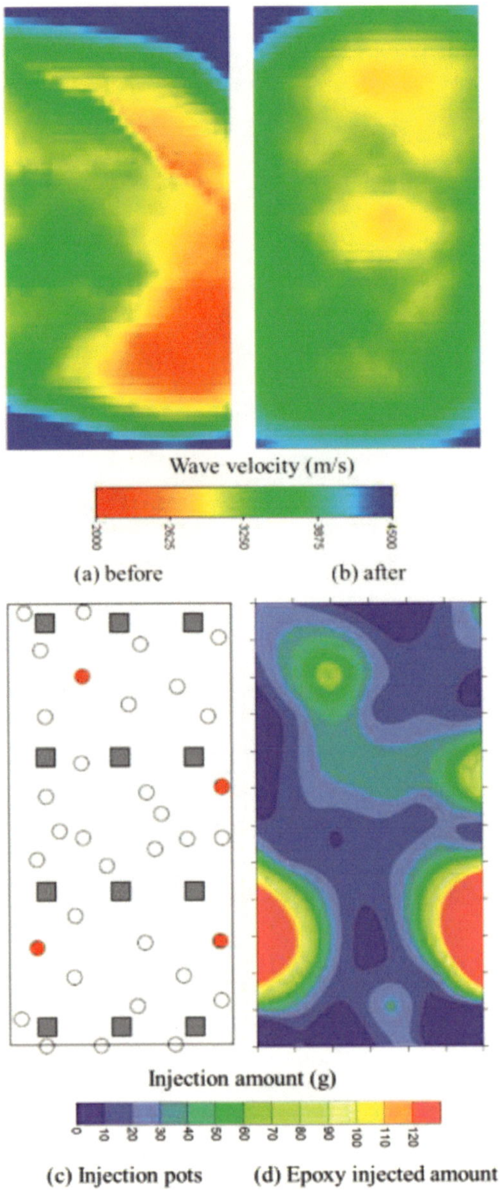

Figure 16.
Wave velocity and epoxy injection result (Side B).

Careful hammering at the steel bar head to prevent contacting the hole wall, the excited elastic waves were generated only from the bit hole bottom into the lining concrete, so that the excited signals were detected finally at sensors located on the concrete surface.

4.2.2 Propagation of waves in steel bar

The travel time along the steel bar was measured by two sensors as shown in **Figure 18**. AE sensor A records the excitation time at the head by using a steel ball of 15 mm diameter and the elastic wave travel time in the bar is calculated by

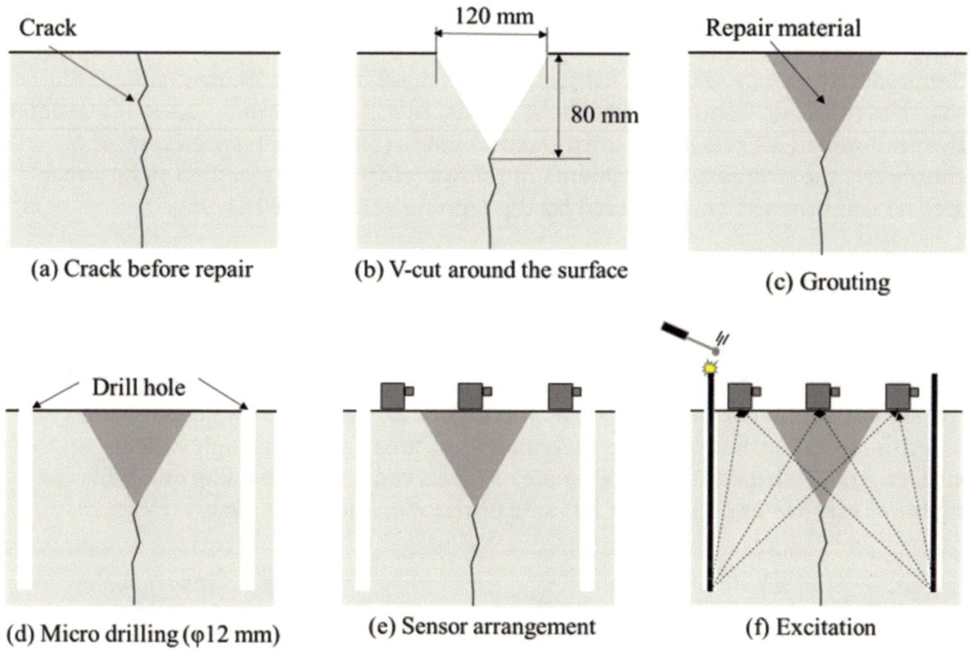

Figure 17.
Procedures of drilling and excitation.

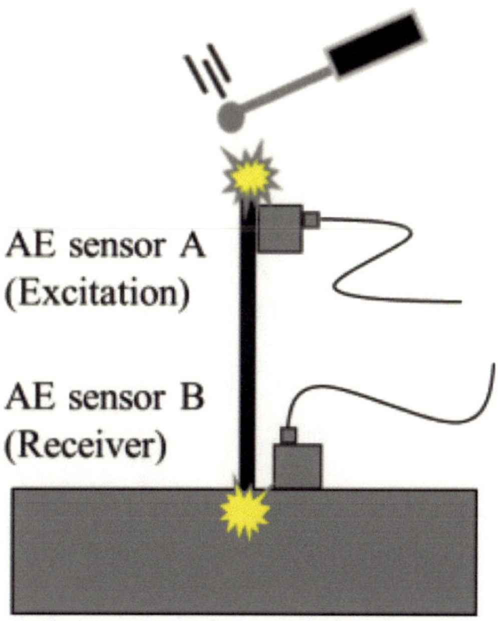

Figure 18.
Measurement method of travel time in the steel bar.

detecting the arrival time of the wave at AE sensor B. The arrival time difference is 69 μs as shown in **Figure 19**.

The dominant frequency of elastic wave excited by a 15 mm diameter steel ball is known as 19.4 kHz according to [12]. Considering a steel bar is used as wave guide, a frequency analysis was conducted for the waveforms observed at A and B.

Figure 20 shows the frequency spectra. Each guided wave is detected via the 38 mm length steel bar. The dominant frequency was observed at A for 22.5 kHz and the dominant frequency was 16.6 kHz at B. Since these detected frequencies are higher than the resonant frequencies of the steel bar, first flexural mode (1.1 kHz), second flexural mode (3.2 kHz), and third flexural mode (5.4 kHz), respectively, as a cantilever, the principal components of the waveform were assumed to be generated as compression wave excited by the tapping at steel bar head.

4.2.3 3D elastic wave tomography

The computation for wave velocity distribution in the targeted concrete wall was implemented by the tomography technique mentioned previously. **Figure 21** shows the 3D distribution of wave velocities and **Figure 22** shows them at cross section A.

Although the triangle-shaped (dashed line) repair area has high velocity on the surface, the V-shaped low-velocity area is observed toward the bottom, whereas high-velocity zones exist at the left side of the specimen. The high velocity may

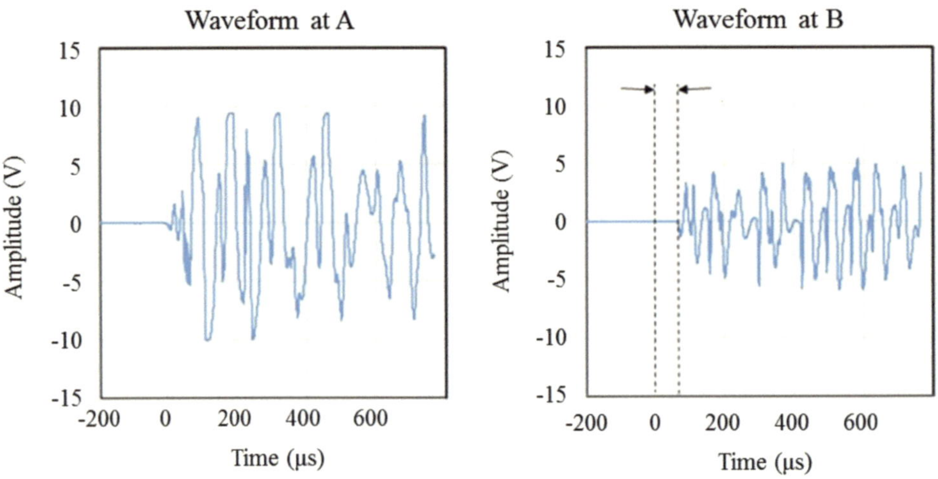

Figure 19.
Waveforms at excitation and receiver.

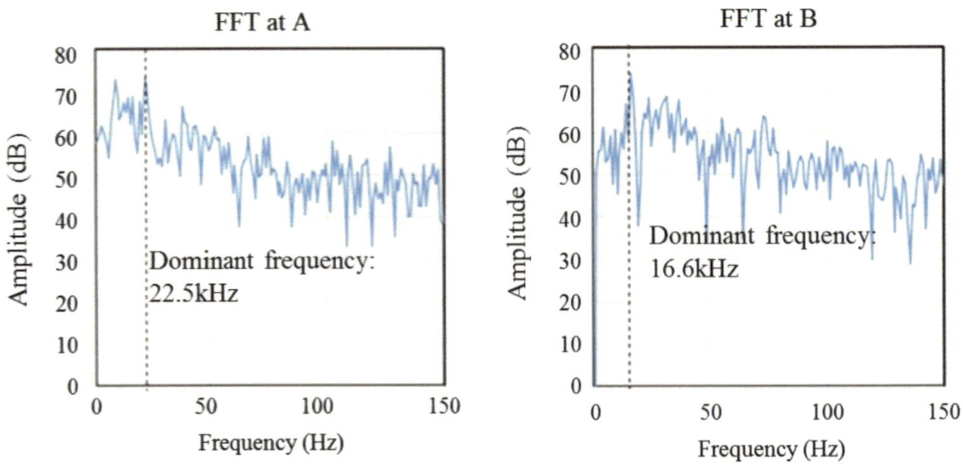

Figure 20.
Frequency spectra of waveforms at A and B.

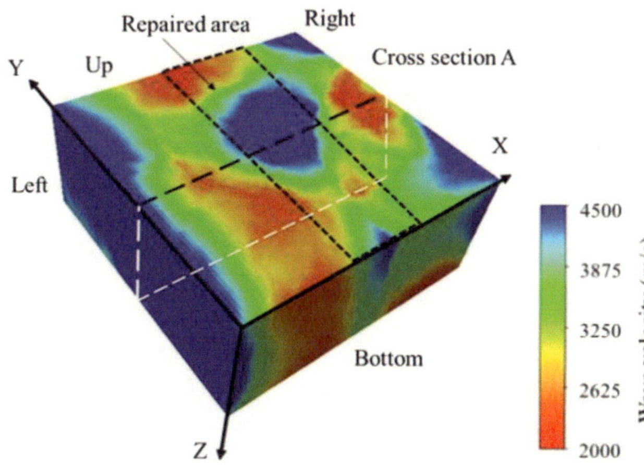

Figure 21.
Distribution of wave velocities in 3D.

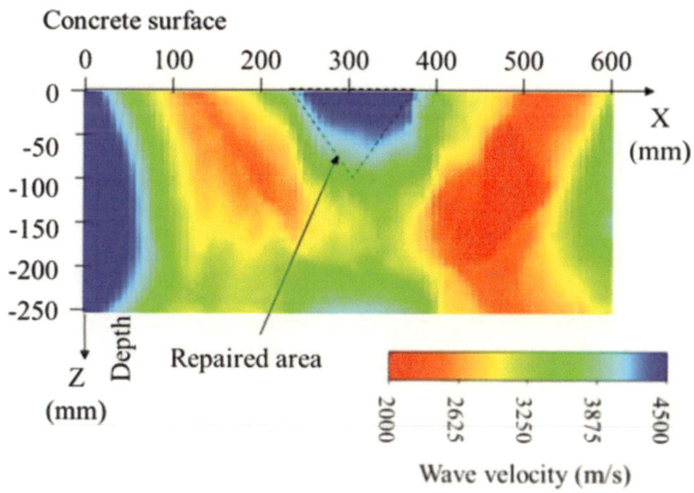

Figure 22.
Distribution of wave velocities at cross section A.

indicate the intact condition of the original concrete quality, because the interested area is enough far from the repaired area.

The repaired part is denoted by high velocity, in **Figure 22**, meanwhile the original concrete surrounding the patched area remarkably shows low velocity. The V-shaped area with low velocity underneath the repaired part could be potentially damaged by the chipping work for concrete removal. This is generally known and described in concrete surface treatment guideline prior to repairs and overlays [13, 14]. Further investigation is needed for the consideration in the influence of the hammer drill impact on damage to the concrete behind the removal zone.

4.3 Slab

4.3.1 Velocity distribution of AE tomography

Figure 23 shows the results of AE tomography, before and after repair by means of the crack injection. Results show that in all the slab panels, the velocity after

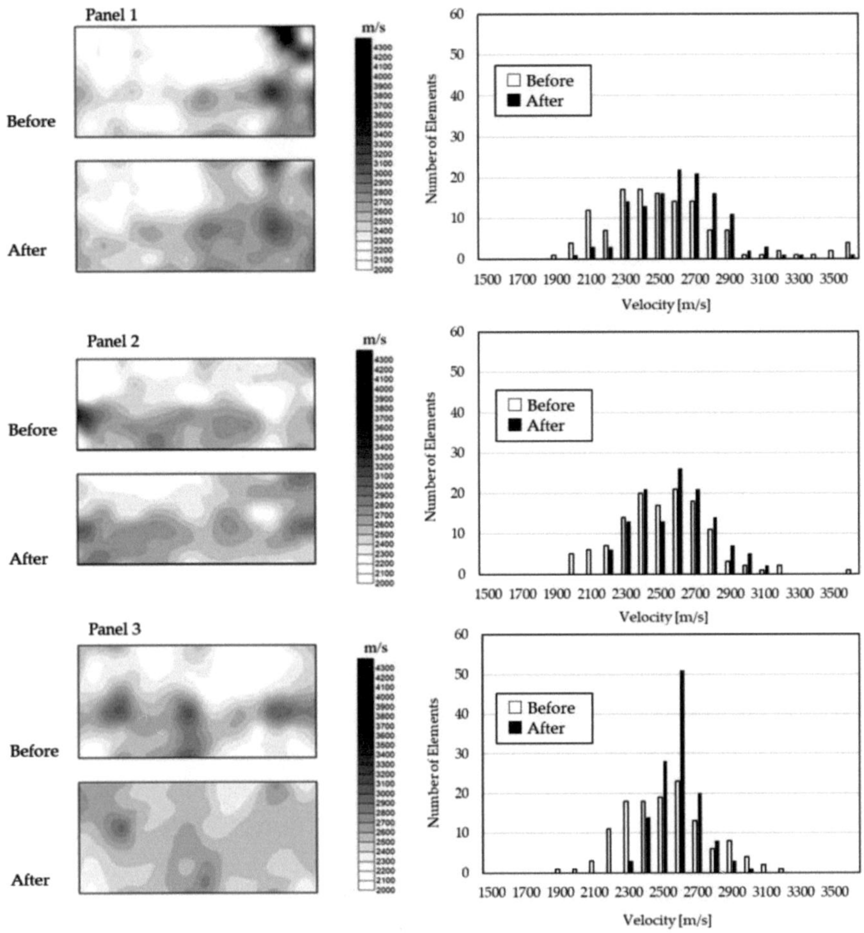

Figure 23.
Results of velocity distribution before and after repair.

repair exhibits increase compared to that before repair. Further quantitatively, the histograms of velocities obtained in all the elements are shown on the right side of the figure. For all slab panels, it is evident that the velocities at the elements clearly shift to the higher regions after repair. Due to the effectiveness of injected material in filling cracks and defects, detours and dispersions in the propagation paths of elastic waves are so eliminated that apparent velocities are increased.

All results imply that the velocity distribution obtained by the AE tomography method has a good potential to be an indicator for ascertaining the filled situation of injected material in a concrete slab. It is confirmed that the velocity for concrete which is not damaged shows about 3500 m/s to 4000 m/s. In some areas, however, velocities of about 2600 m/s are observed even after repair. This is because injected material might not be injected well into continuous cracks, independent air bubbles could be present due to the use of the air-entraining agent, and fine cracks at the interface between coarse aggregate are nucleated due to the alkali-silica reaction. As a result, there exists a possibility that the velocity recovery does not reach to the satisfactory level even after injection. On this issue, we plan to carry out a material test in the laboratory for confirmation.

4.3.2 Relationship between velocity and injection amounts

Design of the injection amount for the crack injection method could be based on the estimation of the crack widths, the depths, and the length measured. It is recognized that there exists no reasonable relationship between the amounts of designed injection and actual injection. Thus, an attempt to examine the amount of injected material is made from the results of AE tomography before repair.

It is considered that the amount of injection should increase, depending on the extent of damage. Namely, if the degree of damage is small, the amount should decrease. In addition, if the damage is less than a certain degree, the injected material may not work well on the damage. On the other hand, if the elastic wave velocity could reflect the degree of damage, a correlation should be evident between

V_P (m/s)	Quality
>4570	Excellent
3660–4570	Fine
3050–3660	Acceptable
2130–3050	Unacceptable
<2130	Poor

Table 2.
Quality indicator (Whitehurst).

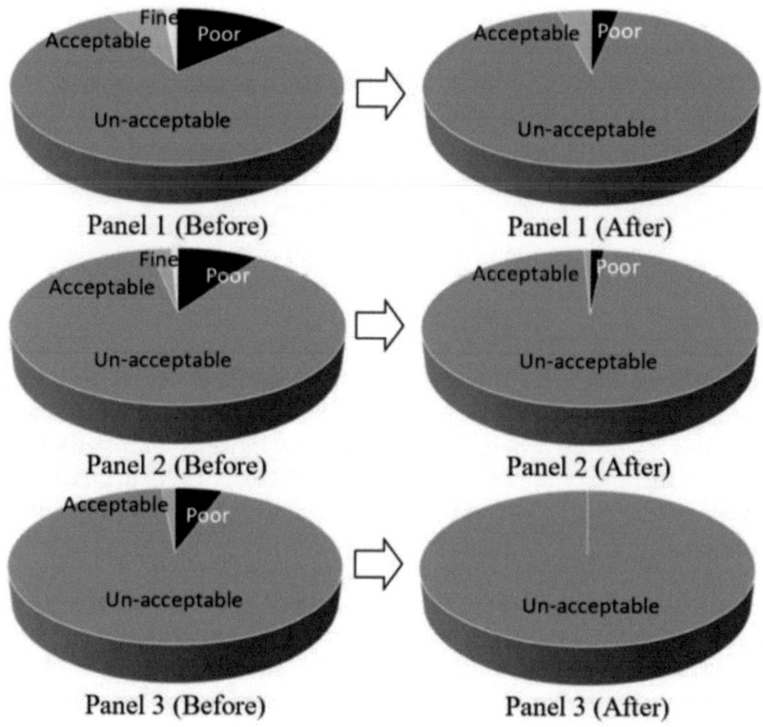

Figure 24.
Area ratio by quality before and after repair.

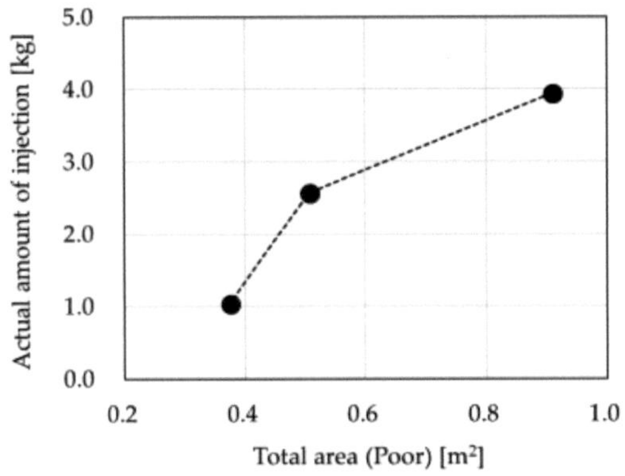

Figure 25.
Total area (Poor) vs. injection amount.

the amount of injection and the values of velocities. Thus, the velocities are classified into grades, as given in **Table 2**. These quality indicators are proposed by Whitehurst [15]. They were determined from the relationship between mechanical properties and P-wave velocity in concrete. Following these indicators, the qualities before and after repair of the panels are classified as shown in **Figure 24**. It is found that the number of elements with Poor decreases, while that of Unacceptable keeps almost the same from before to after repair. As discussed before, due to the presence of air bubbles and the damaged interface with aggregates by alkali-aggregate reaction, the recovery of the velocities may not be apparent. These results imply that the region where the injected material could improve the quality of concrete is mostly that of Poor. It suggests that the repair by means of injection is effective for comparatively major damage. **Figure 25** shows the relationship between total area of Poor estimated by AE tomography before repair and the actual amount of injection. As the Poor area increases, the increase in the actual amount of injection is clearly observed. Thus, it is possible to estimate the amount of injection before repair by carrying out the analysis using AE tomography.

5. Conclusion

Concrete pier, concrete wall, and slab were tested on the investigation on the internal damage assessment for the repair condition by applying elastic wave tomography and AE tomography. Determining the 3D velocity distribution, the repair effects of the epoxy injection method and the patch repair method were quantitatively evaluated. From the results, the following conclusions can be drawn in this study:

1. 3D elastic wave tomography technique can evaluate the penetration of repair epoxy injection material and qualify the repair effect with the amount of injected rexin. 3D tomography technique installed with single-side access drill hammering successfully visualizes the internal quality of concrete after the patch repair method based on the elastic wave velocity distribution.

2. The velocity distribution obtained by AE tomography can serve as an indicator for ascertaining the state of crack and void filling with injected material. A good correlation is found between the low velocity region before repair and the

amount of injected material. The results clearly show the potential for the AE tomography technique to be used as a method for estimating the performance of the crack injection method.

As mentioned previously, the RILEM committee was launched because innovative nondestructive inspection testing to qualify repair works is strongly required worldwide. We plan to continue studies based on the evaluation method using elastic wave tomography and accelerate its standardization.

Author details

Katsufumi Hashimoto[1*], Tomoki Shiotani[1], Takahiro Nishida[1] and Nobuhiro Okude[1,2]

1 Graduate School of Engineering, Kyoto University, Nishikyo-ku, Kyoto, Japan

2 Tokai Technology Center, Higashi-ku, Nagoya, Aichi, Japan

*Address all correspondence to: hashimoto.katsufumi.8a@kyoto-u.ac.jp

References

[1] Damage assessment in Consideration of Repair/Retrofit-Recovery in Concrete and Masonry Structures by Means of Innovative NDT, Technical Committee IAM, RILEM; http://www.rilem.org/gene/main.php?base=8750&gp_id=347

[2] Kobayashi Y, Shiotani T, Shiojiri H. Damage identification using seismic travel time tomography on the basis of evolutional wave velocity distribution model. In: Structural Faults and Repair 2006 (CD-ROM); 2006

[3] Kobayashi Y, Shiotani T, Aggelis DG, Shiojiri H. Three-dimensional seismic tomography for existing concrete structures. In: Proceedings of Second International Operational Analysis Conference; 2007. Vol. 2. pp. 595-600

[4] Kobayashi Y, Shiotani T. Seismic tomography with estimation of source location for concrete structure. In: Structural Faults and Repair 2012, CD-ROM; 2012

[5] Asuae H, Shiotani T, Nishida T, Watabe K, Miyata H. Applicability of AE tomography for accurate damage evaluation in actual RC bridge deck. In: Structural Faults & Repair Conference, No.1743; 2016

[6] Kobayashi Y, Shiotani T. Computerized AE tomography, innovative AE and NDT techniques for on-site measurement of concrete and masonry structures. In: State-of-the-Art Report of the RILEM Technical Committee 239-MCM; Springer: 2016. pp. 47-68

[7] Akaike H. Markovian representation of stochastic processes and its application to the analysis of autoregressive moving average processes. Annals of the Institute of Statistical Mathematics. 1974;**26**(1): 363-387

[8] Zhang H, Thurber C, Rowe C. Automatic P-wave arrival detection and picking with multiscale wavelet analysis for single-component recordings. Bulletin of the Seismological Society of America. 2003;**93**(5):1904-1912

[9] Osawa S, Shiotani T, Kitora H, Momiyama Y. Damage visualization of imperfectly-grouted sheath in PC structures. In: 31st Conference of the European Working Group on Acoustic Emission, German Society for Non-Destructive; 2014. http://www.ewgae2014.com/portals/131/bb/fr1b3.pdf

[10] Sassa K, Ashida Y, Kozawa T, Yamada M. Improvement in the accuracy of seismic tomography by use of an effective ray-tracing algorithm. In: MMIJ/IMM Joint Symposium Volume Papers; 1989. pp. 129-136

[11] Kikusui Chemical Industries Co. Ltd. Inside Pressure Hardening. http://www.kikusui-chem.co.jp/pdf/products/catalog/cat_iph_3.pdf

[12] Sansalone MJ, Streett WB. Impact-Echo, Nondestructive Evaluation of Concrete and Masonry. Ithaca, NY: Bullbrier Press; 1997. pp. 29-46

[13] U.S. Department of the Interior Bureau of Reclamation Technical Service Center. Best Practices for Preparing Concrete Surfaces Prior to Repairs and Overlays; 2012

[14] American Concrete Pavement Association, Guideline for Partial-Depth Spall Repair, Concrete Paving Technlogy; 1998

[15] Whitehurst EA. Evaluation of concrete properties from sonic tests. In: ACI Monograph No.2; ACI; 1966

Supporting open minds since 2005

Neuroprotection
http://dx.doi.org/10.5772/intechopen.77296
Edited by Raymond Chuen-Chung Chang and Yuen-Shan Ho

Contributors
Patrick Flood, Monika Sharma, Maria Bogdan, Liliana Mititelu-Tartau, Victor Gheorman, Gratiela Eliza Popa, Gabriela Rusu, Beatrice Buca, Liliana Pavel, Ina Cristofor, Liliana Foia, Ancuta Goriuc, Cosmin-Gabriel Tartau, Amy Lo, Shen Nian, Arry Yanuar, Widya Dwi Aryati, Rezi Riadhi Syahdi, Nabilah Nurtika, Raymond Chuen-Chung Chang, Yuen-Shan Ho

First published in London, United Kingdom, 2019 by IntechOpen
IntechOpen is the global imprint of INTECHOPEN LIMITED, registered in England and Wales, registration number: 11086078, The Shard, 25th floor, 32 London Bridge Street
London, SE19SG – United Kingdom
Printed in Croatia

British Library Cataloguing-in-Publication Data
A catalogue record for this book is available from the British Library

Additional hard and PDF copies can be obtained from orders@intechopen.com

Neuroprotection
Edited by Raymond Chuen-Chung Chang and Yuen-Shan Ho
p. cm.
Print ISBN 978-1-78984-736-9
Online ISBN 978-1-78984-737-6
eBook (PDF) ISBN 978-1-78985-925-6

For EU product safety concerns:
IN TECH d.o.o., Prolaz Marije Krucifikse Kozulić 3, 51000 Rijeka, Croatia, info@intechopen.com or visit our website at intechopen.com.

Meet the editor

Dr. Chang is the Lab Chief for the Laboratory of Neurodegenerative Diseases in the School of Biomedical Sciences, LKS Faculty of Medicine, the University of Hong Kong. Dr. Chang has organized the International Alzheimer's Disease Conference every year since 2000. He has published over 142 peer-reviewed papers, 14 book chapters, and edited 3 books on neurodegenerative diseases. Dr. Chang is the chief editor for the *American Journal of Alzheimer's Disease and Other Dementias*, senior editor for the *Journal of Neuroimmune Pharmacology*, and associate handling editor for *Frontiers in Neurology/Neurosciences/Psychiatry*. He is on the Scientific Advisory Board of the International AD/PD Symposium, a member of the editorial boards of more than 20 different journals, and a grant reviewer for different grant agencies/foundations.

Dr. Ho has been working for some time on aging-associated neurodegenerative diseases. She is a registered Chinese medicine practitioner with rich experience in both laboratory and clinical research. She has worked as an assistant professor in Macau and is now at the School of Nursing in Hong Kong Polytechnic University. Her research interest includes the use of Chinese herbal medicine and acupuncture to prevent neurodegeneration (neuroprotection). Dr. Ho focuses her research on topics including, but not limited to, disease progression and risk factors leading to neurodegenerative diseases. She has expertise in combining Western and Chinese medicine to elicit holistic effects on the body to exert neuroprotection. She is a member of the editorial boards of four journals and a constant grant reviewer for different journals and funding agencies.

Contents

Preface XI

Section 1
Neurotransmitter Receptor as Target for Neuroprotection 1

Chapter 1 3
Introductory Chapter: Concept of Neuroprotection - A New Perspective
by Raymond Chuen-Chung Chang and Yuen-Shan Ho

Chapter 2 13
Current Therapeutic Approaches from Imidazoline and Opioid Receptors Modulators in Neuroprotection
by Liliana Mititelu-Tartau, Maria Bogdan, Victor Gheorman, Liliana Foia, Ancuta Goriuc, Gabriela Rusu, Beatrice Buca, Liliana Pavel, Ana Cristofor, Cosmin-Gabriel Tartau and Gratiela Eliza Popa

Chapter 3 31
The Role and Development of the Antagonist of Adenosine A_{2A} in Parkinson's Disease
by Widya Dwi Aryati, Nabilah Nurtika Salamah, Rezi Riadhi Syahdi and Arry Yanuar

Chapter 4 45
Adrenergic Receptors as Pharmacological Targets for Neuroinflammation and Neurodegeneration in Parkinson's Disease
by Monika Sharma and Patrick M. Flood

Section 2
Neuroprotection to the CNS 69

Chapter 5 71
Protecting the Aging Retina
by Shen Nian and Amy C.Y. Lo

Preface

Neuroprotection is a key approach to maintaining brain health to prevent the retraction of synapses, accumulation of neurodegenerative proteins (tau, β-amyloid peptide, α-synuclein, huntingtin, or any misfolded proteins), activation of neuroinflammation, or even the breakdown of the blood-brain barrier. Neuroprotection can be applied to chronic neurodegenerative diseases (Alzheimer's disease, Parkinson's disease, Lewy's body diseases, Huntington's disease, amyotrophic lateral sclerosis, or even brain tumors), acute neurological disorders (brain trauma, stroke, transient ischemia, epilepsy, or cerebral hemorrhage), or even mental disorders (psychosis, depression, or schizophrenia). The traditional view of neuroprotection is often restricted to the use of agonists or antagonists of neurotransmitter receptors. While this is still a great approach to safeguard neurons, neuroprotection strategy is not limited to modulating neurotransmitter receptors. Neuroprotection can be accomplished by (1) natural products, (2) herbal medicine, (3) different forms of exercise, (4) learning new skills or languages, or (5) even good sleeping patterns. From all these new perspectives, the concept of neuroprotection is limited not only to the protection of neurons in the pathogenesis of diseases but also to the prevention of any detrimental factors leading to neuronal cell loss.

The major aim of this book is to focus on different approaches to achieve neuroprotection. In this book, most of our authors review the advantages of neurotransmitter receptors. Mititelu-Tartau and Bogdan's group reviews the imidazoline ligands and opioid ligands in Alzheimer's disease. They review the neuroprotective effects of agmatine on memory. In addition, they also summarize the findings of using different opioid receptor ligands to elicit neuroprotection. Similarly, Yanuar's group reviews the beneficial effects of the adenosine A_{2A} receptor antagonist. It has been reported that the adenosine receptor antagonist elicits neuroprotective effects, partially because of its signaling pathways to activate the cytosolic fraction of cAMP. Yanuar's group reviews how A_{2A} antagonists provide neuroprotection.

Apart from modulating neurotransmitter receptors, Flood's group reviews another very important pathological factor leading to neurodegeneration: neuroinflammation. Microglial cells are the major line of cells to be activated to produce cytokines and free radicals to damage neurons. Interestingly, Flood's group reviews how different adrenergic receptor agonists and antagonists modulate microglial responses. This independent review suggests that modulation of neuroinflammation can also be achieved by intervening neurotransmitter receptors and their associated signaling events.

Neuroprotection can be disease specific but use different approaches. Nian and Lo review the neuroprotection of aging eyes, in which aging-associated macular degeneration often occurs in the elderly. With the progression of the disease and identification of biological targets at different states of the disease, the strategy of neuroprotection can be changed.

Taken together, neuroprotection receives increasing attention from different approaches in different states of neurodegenerative diseases, acute neurological disorders, and even mental disorders. Even in disease progression, neuroprotection can prevent the spread of neuronal damage or neuronal loss from one region to other regions. This book will give insights to scientists in the field to stimulate their research, medical professionals to review their clinical practices, and others who would like to learn more about different neuroprotective approaches.

Raymond Chuen-Chung Chang, PhD
Laboratory of Neurodegenerative Diseases,
School of Biomedical Sciences,
LKS Faculty of Medicine,
The University of Hong Kong,
Pokfulam, Hong Kong

Yuen-Shan Ho, PhD
School of Nursing,
Faculty of Health and Social Sciences,
The Hong Kong Polytechnic University,
Kowloon, Hong Kong

Section 1

Neurotransmitter Receptor as Target for Neuroprotection

Chapter 1

Introductory Chapter: Concept of Neuroprotection - A New Perspective

Raymond Chuen-Chung Chang and Yuen-Shan Ho

1. Introduction

Neuroprotection is an approach to preserve neurons so that neurons cannot be hurt by different pathological factors in neurodegenerative diseases. It can be an approach before the onset of the disease so that neurons cannot be affected by any risk factors. It can also be an approach during the progression of the disease to prevent spreading of injury from one neuron to neighboring neurons. Therefore, neuroprotection can also be one approach as "disease-modifying agent" to delay and even stop progressive neurodegeneration.

The concept of neuroprotection has long been confined to intervene neurotransmitter receptors via agonists and antagonists. A very well-known example is the neuroprotective effect of caffeine via adenosine receptor, because caffeine is an A2 receptor antagonist. Caffeine can exert neuroprotection via adenosine A2 receptor to protect dopaminergic neurons in an experimental model of Parkinson's disease (PD) by using 1-methyl-4-phenyl-1,2,3,6-tetrahydropyridine (MPTP) as a toxin agent [1]. Therefore, drinking coffee has been considered to be neuroprotection. Apart from adenosine receptor on neurons, it is now known that microglial cells also express A2 receptor, and A2 receptor antagonist or caffeine can reduce activation of microglial cells [2]. Since caffeine or A2 receptor antagonists can protect neurons and minimize activation of microglial cells, research is still very active in this direction. Indeed, investigation of other neurotransmitter receptor antagonists, agonists, blockers, or even partial blockers is still a very active research area in neuroprotection.

While it is exciting to reveal differential neuroprotective effects via modulating receptors for different neurotransmitters, we should not restrict ourselves to this perspective only. In fact, a wider scope of neuroprotection has been evolved. The concept of neuroprotection can now be categorized into three groups: (A) pharmacological intervention, (B) non-pharmacological intervention, and (C) cellular and genetic approaches (**Figure 1**).

In group A neuroprotective approaches, pharmacological intervention includes modulation of neurotransmitter receptors (as introduced above), anti-oxidative stress, and anti-inflammatory responses [3]. These are classical pharmacological approaches. These methods have been explored for decades and can be used for both neuroprotection and as disease-modifying agents. Their effects can be target-specific for one single protein (e.g., receptor for neurotransmitter or one type of cytokines) or pathway-specific (anti-oxidative responses using nuclear factor erythroid 2-related factor 2, Nrf2). Since single biological target may limit the beneficial effects and may not be able to intervene the complexity of the disease

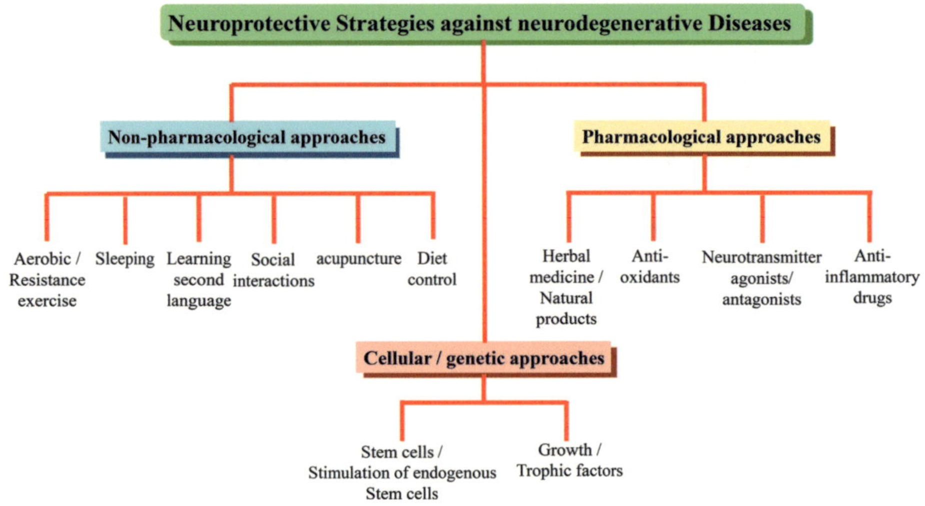

Figure 1.
A summary of different neuroprotective approaches.

progression, neuroprotective effects from herbal medicine or natural products receive increasing attention in scientific research because of their multiple effects on different biological targets [4–6]. For example, it has been shown that triterpenoids extracted from *Prunella vulgaris* has anti-inflammatory effects [7]. We have proved that polysaccharide part of *Prunella vulgaris* can modulate the immune responses of microglia and macrophages [8, 9]. Therefore, *Prunella vulgaris* exerts neuroprotection toward neurodegenerative diseases against neuroinflammation.

In group B neuroprotective approaches, non-pharmacological approaches receive increasing attention. Since it is quite difficult to ask a healthy adult or even elderly to constantly take medicine even though it is herbal, non-pharmacological approaches to earn neuroprotection are most welcome. Among these different methods, exercise is the best to prevent nearly all kinds of diseases. Exercise can be further divided into aerobic and resistance exercises, in which aerobic exercise is suitable for motor-healthy adults or elderly. However, there is a large population of elderly who have difficulty in their motor functions. Resistance exercise becomes an excellent tool for them. Both aerobic and resistance exercises can affect the metabolisms of the body and induce trophic factors like myokines, released from muscles and liver [10, 11]. Since body metabolism is important, diet control is also important to reduce any risk factors leading to the development of neurodegenerative diseases, such as diabetes, hyperlipidemia, or hypercholesterolemia. All of these risk factors are related to vascular components, which may further add on vascular dementia [12–14]. Apart from exercise, it is well-known that, maintaining social activity is essential to keep our brain healthy [15]. In addition, sleeping can promote clearing of pathological factors, such as β-amyloid (Aβ) peptide and phosphorylated tau protein [16, 17]. Also, it has been recently shown that learning second language or bilingualism can be neuroprotective [18]. Furthermore, increasing lines of evidence have shown that acupuncture can help adjusting the body metabolism and immunity [18, 19]. Therefore, acupuncture can also be considered to be one method in non-pharmacological approaches. One key point should be noticed that all these neuroprotective approaches are multiple targets. This may be why they are so effective in minimizing neurodegeneration to preserve neurons.

Neuroprotective approaches in group C are the new extension of multiple effects from trophic factors secreted by genetically engineered cells or viral vector. Alternatively, stimulating the proliferation and differentiation of endogenous stem cells, application of induced pluripotent stem cells (iPSC), or mesenchymal stem cells are popular trends in scientific research to prevent neurodegeneration and neuronal loss [20–22].

Since there are several chapters about the effects of neurotransmitter receptors, we will focus on some neuroprotective approaches only, as this is an introductory chapter.

2. Neuroprotective effects of herbal medicine and natural products

This direction of neuroprotection is the most controversial approach among different pharmacological tools. This is because most of the herbal medicine or natural products have a wide array of effects. If we use traditional way of thinking to target a specific protein or signaling pathway, herbal medicine fails to do so. However, if we are aware of neurodegenerative diseases that are usually multifactorial, we can then accept that herbal medicine should be the direction to prevent neuronal loss and spreading of neurodegeneration. Our group has long been working on discovering herbal medicine as neuroprotective agents and investigating their underlying mechanisms. Since most of the herbal medicines are in decoction form, for oral consumption, the effective components in different herbs are not limited to one single chemical. The small molecules being extracted from one herb can also be found from other herbs. There are many laboratories investigating those small molecules and have Research and Development (R&D) to be commercial products. In contrast, big molecules such as polysaccharides have been an unexplored area. In fact, a large portion of components from herbs is polysaccharides. When we use hot water to prepare decoction, what we can easily absorb is polysaccharides. Therefore, we have investigated polysaccharides extracted from different herbs [23].

We have first found that polysaccharides from *Nerium indicum* exert neuroprotective effects against β-amyloid (Aβ) neurotoxicity cultured neurons [24, 25]. Then, we also found that neuroprotective effects of *Verbena officinalis* and *Ganoderma lucidum* [26, 27]. Having investigated different herbal medicine, we found that the polysaccharides from *Lycium barbarum* (Wolfberry) are potent [28, 29]. Polysaccharides can be extracted in hot water or from alkaline condition and elicit neuroprotection [30]. It should be noted that not all polysaccharides, but only some sub-fractions provide neuroprotective effects. Apart from neurons in the brain, polysaccharides from wolfberry can also protect the retina against experimental glaucoma and stroke [31–33]. As harvesting and planting processes have good agricultural practice and the extraction can be done under stringent control of good manufacturing practice, we choose wolfberry as our long-term study. Wolfberry is indeed an anti-aging Chinese herbal medicine. We would like to understand the concept of "anti-aging." Therefore, we have investigated its application in different aging-associated neurodegenerative diseases [34]. In fact, not all aging-associated neurodegenerative diseases can be attenuated by the polysaccharide fraction of Wolfberry. In Parkinson's disease (PD), more potent antioxidants are required to provide neuroprotection.

In experimental PD using 6-hydroxydopamine (6-OHDA) as toxin agent, we employed oxyresveratrol, which is a natural product but a structural analog of resveratrol, and found that it can attenuate neurodegeneration [35, 36]. For this kind of neurodegenerative disease requiring high levels and high capacity of antioxidant, another method is to employ pro-drug approach so that oxidant cleave the precursor and the product becomes potent antioxidant. Interestingly, polyphenol (-)-epigallocatechin-3-gallate (EGCG) from green tea falls in this category [37].

The advantages of using herbal medicine or natural products are to make use of their multiple effects. In neurodegenerative diseases, a wide array of stress responses is stimulated because of free radicals or accumulation of misfolded or badly folded proteins. Most of the herbal medicine and natural products attenuate many stress kinases. They may not be a good candidate to clear accumulation of bad proteins; however, they can inhibit the cascades of stress responses (e.g., activation of c-Jun-N-terminal kinase, JNK; endoplasmic reticulum stress pathways), which usually lead to neuronal apoptosis. Some of the herbal medicine or natural products can even strengthen the survival signaling pathways such as mTOR or Akt pathways.

In addition to their direct effects on neurons, there are many herbal medicine or natural products that can modulate body immune responses. For example, polysaccharides fraction from *Prunella vulgaris* L. can modulate innate immune responses of macrophages [8, 9]. It has been increasingly aware that activation of body (systemic) immune responses are the origin of our sickness responses leading to low appetites and even fever [38]. Long-term effects of systemic immune responses can even result in psychological depression or acute delirium [39]. Experimentally, we have shown that infection in the body or immune responses triggered by surgery (laparotomy) can modulate cognitive functions because of stimulation of neuroimmune responses [40, 41]. Therefore, modulation of body immunity can be a powerful method to minimize neuroimmune responses and then reduce cognitive dysfunctions.

3. Aerobic and resistance exercises

As emphasized above, the advantages of herbal medicine and natural products are their multiples and wide array effects. Similarly, exercise is another method to achieve this goal. Exercise can stimulate production of different trophic/growth factors, e.g., vascular endothelial growth factor (VEGF), brain-derived neurotrophic factor (BDNF), insulin-like growth factor-1 (IGF-1), peroxisome proliferator-activated receptor gamma coactivator-1α (PGC-1α)/neuronal fibronectin type III domain-containing protein 5 (FNDC5). Some trophic factors can be secreted by muscles or liver, which is collectively called myokines. It has been shown that muscle cells can produce PGC-1α, irisin, cathepsin B, myostatin (growth differentiation factor 8), fibroblast growth factor 21 (FGF-21), IL-6, IL-15, myonectin, and SPARC (osteonectin) as novel brain-beneficial myokines [10, 42]. Some of them, such as FGF21, can also be secreted by the liver or even in the brain upon exercise [43]. They all can affect muscle, liver, adipose tissues, and brain cells, leading to preservation of cognitive functions, modulating the lipid metabolism, and immune responses [44]. This kind of myokine can pass through the blood-brain barrier.

For motor-healthy subject, aerobic exercise is beneficial to majority of people. As discussed above, myokines can be induced so that they provide neuroprotective effects to the brain and attenuate any systemic immune responses. While aerobic exercise is known to be good, a great number of elderly and demented patients could not enjoy the benefit of aerobic exercise. Those patients may have prior knee replacement or fall down hurting the legs. They may be lying down on the bed. No matter how we advocate the beneficial effects of aerobic exercise, this will not be beneficial to this group of patients. Does it mean that they will not have any beneficial effects of exercise? This question leads us to investigate the beneficial effects of resistance exercise.

In a systematic review, aerobic exercise shows beneficial effects on cognitive functions and executive functions, which is better than that from resistance exercise [45]. Furthermore, aerobic exercise seems to elicit strong anti-inflammatory effect than that of resistance exercise [46]. However, resistance exercise seems to particularly increase the volume of hippocampus [47]. A study in laboratory animal has shown that aerobic exercise can markedly increase BDNF; whereas, resistance exercise can significantly increase IGF-1 [48]. Although resistance exercise may have some limitations, this is still a good choice for those elderly and dementia patients to have this form of exercise. IGF-1 can still elicit multiple functions in the brain to preserve neurons.

4. Concluding remarks

Since this is an introductory chapter, we do not intend to discuss every single method of neuroprotection in detail. We should keep in mind that neurodegenerative diseases are multiple-hit processes. Therefore, no single biological target can afford all neuroprotective needs. Approaches in using herbal medicine and natural products remind us that multiple biological targets may be the way to exert effective neuroprotection. Therefore, any non-pharmacological approaches including exercise and even the new stem cell approaches exert multiple effects. This can be the way for our effective neuroprotective strategies.

Author details

Raymond Chuen-Chung Chang[1,2]* and Yuen-Shan Ho[3]

1 Laboratory of Neurodegenerative Diseases, LKS Faculty of Medicine, The University of Hong Kong, Pokfulam, Hong Kong SAR

2 State Key Laboratory of Brain and Cognitive Sciences, The University of Hong Kong, Pokfulam, Hong Kong SAR

3 Faculty of Health and Social Sciences, School of Nursing, The Hong Kong Polytechnic University, Kowloon, Hong Kong SAR

*Address all correspondence to: rccchang@hku.hk

References

[1] Chen JF, Xu K, Petzer JP, et al. Neuroprotection by caffeine and A(2A) adenosine receptor inactivation in a model of Parkinson's disease. The Journal of Neuroscience. 2001;**21**:RC143

[2] Caetano L, Pinheiro H, Patrício P, et al. Adenosine A2A receptor regulation of microglia morphological remodeling-gender bias in physiology and in a model of chronic anxiety. Molecular Psychiatry. 2017;**22**:1035-1043

[3] Chang RCC, Chiu K, Ho YS, So KF. Modulation of neuroimmune responses on glia in the central nervous system: Implication in therapeutic intervention against neuroinflammation. Cellular & Molecular Immunology. 2009;**6**:317-326

[4] Ho YS, So KF, Chang RCC. Anti-aging herbal medicine—How and why can they be used in aging-associated neurodegenerative diseases? Ageing Research Reviews. 2010;**9**:354-362

[5] Ho YS, So KF, Chang RCC. Drug discovery from Chinese medicine against neurodegeneration in Alzheimer's and vascular dementia. Chinese Medicine. 2011;**6**:15. DOI: 10.1186/1749-8546-6-15

[6] Chao J, Leung Y, Wang M, Chang RCC. Nutraceuticals and their preventive or potential therapeutic value for Parkinson's disease. Nutrition Reviews. 2012;**70**:373-386

[7] Wang YY, Yang YX, Zhe H, He ZX, Zhou SF. Bardoxolone methyl (CDDO-me) as a therapeutic agent: An update on its pharmacokinetic and pharmacodynamic properties. Drug Design, Development and Therapy. 2014;**8**:2075-2088

[8] Fang X, Yu MS, Yuen WH, Zee SY, Chang RCC. Immune modulatory effects of *Prunella Valgaris L.* on monocytes/macrophages. International Journal of Molecular Medicine. 2005;**16**:1109-1116

[9] Fang X, RCC C, Yuen WH, Zee SY. Immune modulatory effects of *Prunella Valgaris L*. International Journal of Molecular Medicine. 2005;**15**:491-496

[10] Liu Y, Yan T, Chu JMT, Chen Y, Dunnett S, Ho YS, et al. The beneficial effects of physical exercise in the brain and related pathophysiological mechanisms in neurodegenerative diseases. Laboratory Investigation. 2019. DOI: 10.1038/s41374-019-0232-y. In press

[11] Marston KJ, Brown BM, Rainey-Smith SR, Peiffer JJ. Resistance exercise-induced responses in physiological factors linked with cognitive health. Journal of Alzheimer's Disease. 2019. DOI: 10.3233/JAD-181079. In press

[12] Li G, Mayer CL, Morelli D, et al. Effect of simvastatin on CSF Alzheimer disease biomarkers in cognitively normal adults. Neurology. 2017;**89**:1251-1255

[13] Appleton JP, Scutt P, Sprigg N, Bath PM. Hypercholesterolaemia and vascular dementia. Clinical Science (London, England). 2017;**131**:1561-1578

[14] Besser LM, Alosco ML, Ramirez Gomez L, et al. Late-life vascular risk factors and Alzheimer disease neuropathology in individuals with normal cognition. Journal of Neuropathology and Experimental Neurology. 2016;**75**:955-962

[15] Rege SD, Geetha T, Broderick TL, Babu JR. Can diet and physical activity limit Alzheimer's disease risk? Current Alzheimer Research. 2017;**14**:76-93

[16] Zhang F, Zhong R, Li S, Chang RCC, Le W. The missing link between sleep disorders and age-related dementia: Recent evidence and plausible mechanisms. Journal of Neural Transmission. 2017;**124**:559-568

[17] Cox SR, Bak TH, Allerhand M, Redmond P, Starr JM, Deary IJ, et al. Bilingualism, social cognition and executive functions: A tale of chickens and eggs. Neuropsychologia. 2016;**91**:299-306

[18] Cao Y, Zhang LW, Wang J, Du SQ, Xiao LY, Tu JF, et al. Mechanisms of acupuncture effect on Alzheimer's disease in animal based researches. Current Topics in Medicinal Chemistry. 2016;**16**:574-578

[19] Leung MC, Yip KK, Ho YS, Siu FK, Li WC, Garner B. Mechanisms underlying the effect of acupuncture on cognitive improvement: A systematic review of animal studies. Journal of Neuroimmune Pharmacology. 2014;**9**:492-507

[20] Robbins JP, Price J. Human induced pluripotent stem cells as a research tool in Alzheimer's disease. Psychological Medicine. 2017;**47**:2587-2592

[21] Arber C, Lovejoy C, Wray S. Stem cell models of Alzheimer's disease: Progress and challenges. Alzheimer's Research & Therapy. 2017;**9**:42. DOI: 10.1186/s13195-017-0268-4)

[22] Peng X, Xing P, Li X, Qian Y, Song F, Bai Z, et al. Towards personalized intervention for Alzheimer's disease. Genomics, Proteomics & Bioinformatics. 2016;**14**:289-297

[23] Ho YS, Poon DCH, Chan TF, Chang RCC. From small to big molecules: How do we prevent and delay progression of aging-related neurodegeneration? Current Pharmaceutical Design. 2012;**18**:15-26

[24] Yu MS, Lai SW, Lin KF, Fang JN, Yuen WH, Chang RCC. Characterization of polysaccharides from the flowers of *Nerium indicum* and their neuroprotective effects. International Journal of Molecular Medicine. 2004;**14**:917-924

[25] Yu MS, Wong AYY, So KF, Fang JN, Yuen WH, Chang RCC. New polysaccharide from *Nerium indicum* protects neurons via stress kinase signaling pathway. Brain Research. 2007;**1153**:221-230

[26] Lai SW, Yu MS, Yuen WH, Chang RCC. Novel neuroprotective effects of the aqueous extracts from *Verbena officinalis Linn*. Neuropharmacology. 2006;**50**:641-650

[27] Lai CSW, Yu MS, Yuen WH, So KF, Zee SY, Chang RCC. Neuroprotective effects of anti-aging fungus *Gandoderma lucdium*. Brain Research. 2008;**1190**:215-224

[28] Yu MS, Leung SKY, Che CM, Zee SY, So KF, Yuen WH, et al. Neuroprotective effects of an anti-aging Chinese medicine *Lycium barbarum* against β-amyloid peptide neurotoxicity. Experimental Gerontology. 2005;**40**:716-727

[29] Yu MS, Ho YS, So KF, Yuen WH, Chang RCC. Cytoprotective effects of *Lycium barbarum* against reducing stress on endoplasmic reticulum. International Journal of Molecular Medicine. 2006;**17**:1157-1162

[30] Ho YS, Yu MS, Lai CSW, So KF, Yuen WH, Chang RCC. Characterizing the neuroprotective effects of from *Lycium barbarum* on β-amyloid peptide neurotoxicity. Brain Research. 2007;**1158**:123-134

[31] Chan HC, Chang RCC, Ip AKC, Chiu K, Yuen WH, Zee SY, et al. Neuroprotective effects of *Lycium barbarum Lynn*, a traditional

Chinese herbal medicine in protecting retinal ganglion cells in an ocular hypertension model of glaucoma. Experimental Neurology. 2007;**203**:269-273

[32] Chiu K, Zhou Y, Yeung SC, Lok CKM, Chan OOC, Chang RCC, et al. Up-regulation of crystallins is involved in neuroprotective effects of wolfberry on survival of retinal ganglion cells in rat ocular hypertension model. Journal of Cellular Biochemistry. 2010;**110**:311-320

[33] Li SY, Yang D, Yeung CM, Yu WY, Chang RCC, So KF, et al. *Lycium barbarum* polysaccharides reduce neuronal damage, blood-retinal barrier disruption and oxidative stress in retinal ischemia/reperfusion injury. PLoS One. 2011;**6**:e16380

[34] Chang RCC, So KF. Use of anti-aging herbal medicine, *Lycium barbarum*, against aging-associated diseases. What do we know so far? Cellular and Molecular Neurobiology. 2008;**28**:643-652

[35] Chao J, Yu MS, Ho YS, Wang M, Chang RCC. Dietary oxyresveratrol prevents parkinsonian mimetic 6-hydroxydopamine neurotoxicity. Free Radical Biology & Medicine. 2008;**45**:1019-1026

[36] Chao J, Li H, Cheng KW, Yu MS, Chang RCC, Wang M. Protective effects of pinostilbene, a resveratrol methylated derivative, against 6-hydroxydopamine-induced neurotoxicity in SH-SY5Y cells. The Journal of Nutritional Biochemistry. 2010;**21**:482-489

[37] Chao J, Lau KW, Huie MJ, Ho YS, Yu MS, Lai CSW, et al. A prodrug of the green tea polyphenol (−)-epigallocatechin-3-gallate (EGCG) prevents differentiated SH-SY5Y cells from toxicity induced by 6-hydroxydopamine. Neuroscience Letters. 2010;**469**:360-364

[38] Poon DCH, Ho YS, Chiu K, Chang RCC. Cytokines: How important are they in mediating sickness? Neuroscience and Biobehavioral Reviews. 2013;**37**:1-10

[39] Poon CH, Ho YS, Chiu K, Wong HL, Chang RCC. Sickness: From the focus on cytokines, prostaglandins, and complement factors to the perspectives of neurons. Neuroscience and Biobehavioral Reviews. 2015;**57**:30-45

[40] Poon DCH, Ho YS, You R, Tse J, Chu K, Chang RCC. PKR deficiency alters E. Coli-induced sickness behaviors but does not exacerbate neuroimmune responses or bacterial load. Journal of Neuroinflammation. 2015;**12**:212. DOI: 10.1186/s12974-015-0433-2

[41] Huang CX, Irwin MG, Wong GTC, Chang RCC. Evidence of the impact of systemic inflammation on neuroinflammation from a non-bacterial endotoxin animal model. Journal of Neuroinflammation. 2018;**15**:147. DOI: 10.1186/s12974-018-1163-z

[42] Lee JH, Jun HS. Role of myokines in regulating skeletal muscle mass and function. Frontiers in Physiology. 2018;**10**:42. DOI: 10.3389/fphys.2019.00042

[43] Tanimura Y, Aoi W, Takanami Y, Kawai Y, Mizushima K, Naito Y, et al. Acute exercise increases fibroblast growth factor 21 in metabolic organs and circulation. Physiological Reports. 2016;**4**:e12828. DOI: 10.14814/phy2.12828

[44] Fisher FM, Maratos-Flier E. Understanding the physiology of FGF21. Annual Review of Physiology. 2016;**78**:223-241

[45] Barha CK, Davis JC, Falck RS, Nagamatsu LS, Liu-Ambrose T. Sex differences in exercise efficacy to improve cognition: A systematic review

and meta-analysis of randomized controlled trials in older humans. Frontiers in Neuroendocrinology. 2017;**46**:71-85

[46] King DE, Carek P, Mainous AG 3rd, Pearson WS. Inflammatory markers and exercise: Differences related to exercise type. Medicine & Science in Sports & Exercise. 2003;**35**:575-581

[47] Kim YS, Shin SK, Hong SB, Kim HJ. The effects of strength exercise on hippocampus volume and functional fitness of older women. Experimental Gerontology. 2017;**97**:22-28

[48] Cassilhas RC, Lee KS, Fernandes J, Oliveira MG, Tufik S, Meeusen R, et al. Spatial memory is improved by aerobic and resistance exercise through divergent molecular mechanisms. Neuroscience. 2012;**202**:309-317

Chapter 2

Current Therapeutic Approaches from Imidazoline and Opioid Receptors Modulators in Neuroprotection

Liliana Mititelu-Tartau, Maria Bogdan, Victor Gheorman, Liliana Foia, Ancuta Goriuc, Gabriela Rusu, Beatrice Buca, Liliana Pavel, Ana Cristofor, Cosmin-Gabriel Tartau and Gratiela Eliza Popa

Abstract

Due to brain plasticity, the nervous system is capable of manifesting behavioral variations, adapted to the influences from both external and internal environment. Multiple neurotransmitters are involved in the mediation of pathological processes at the molecular, cellular, regional, and interregional levels participating in cerebral plasticity, their intervention being responsible for various structural, functional, and behavioral disturbances. The current therapeutic strategies in neuroprotection aim at blocking on different levels, the molecular cascades of the pathophysiological mechanisms responsible for neuronal dysfunctions and ultimately for neuronal death. Different agents influencing these neurotransmitters have demonstrated beneficial effects in neurogenesis and neuroprotection, proved in experimental animal models of focal and global ischemic injuries. Serotonin, dopamine, glutamate, N-methyl-D-aspartate, and nitric oxide have been shown to play a significant role in modulating nervous system injuries. The imidazoline system is one of the important systems involved in human brain functioning. Experimental investigations have revealed the cytoprotective effects of imidazoline I2 receptor ligands against neuronal injury induced by hypoxia in experimental animals. The neuroprotective effects were also highlighted for kappa and delta receptors, whose agonists demonstrated the ability to reduce architectural lesions and to recover neuronal functions of animals with experimentally induced brain ischemia.

Keywords: neuroprotection, neurodegenerative diseases, ischemic stroke, imidazoline, opioids, nitric oxide

1. Introduction

Increase in life expectancy has led to aging of the population and consequently to an expansion of the prevalence of neurodegenerative diseases (NDDs) [1].

Clinical type	Disease/disorder
Signs of progressive dementia with no other neurological signs (absent/inconspicuous)	• Alzheimer's disease • Frontotemporal dementias • Some cases of Lewy-body disease • Posterior cortical atrophy (visuospatial dementia)
Signs of progressive dementia accompanied by other neurological abnormalities	• Huntington's disease (chorea) • Lewy-body disease (Parkinsonian features) • Some cases of Parkinson's disease • Corticobasal ganglionic degeneration • Cortical-striatal-spinal degeneration (Jakob's disease) • Dementia-Parkinson-amyotrophic lateral sclerosis complex • Cerebrocerebellar degeneration • Familial dementia with spastic paraparesis, amyotrophy, or myoclonus • Polyglucosan body disease • Frontotemporal dementia with parkinsonism or ALS
Signs of movement disorders or other posture abnormalities	• Parkinson's disease • Multiple system atrophy • Essential tremor • Progressive supranuclear palsy • Dystonia musculorum deformans • Huntington's disease (chorea) • Acanthocytosis with chorea • Corticobasal ganglionic degeneration • Lewy-body disease • Restricted dystonia
Signs of progressive ataxia	• Spinocerebellar ataxias • Cerebellar cortical ataxias • Complicated hereditary and sporadic cerebellar ataxias
Signs of slowly developing muscular weakness and atrophy	• Motor disorders with amyotrophy • Spastic paraplegia without amyotrophy
Sensory and sensorimotor disorders	• Hereditary sensorimotor neuropathies • Pure or predominantly sensory or motor neuropathic • Riley-Day autonomic degeneration
Signs of progressive blindness with or without other neurological disorders	• Pigmentary degeneration of retina • Stargardt's disease • Age-related macular degeneration
Signs characterized by degenerative neurosensory deafness	• Pure neurosensory deafness • Hereditary hearing loss with retinal diseases • Hereditary hearing loss with system atrophies of the nervous system

**Adapted from Ropper A, Samuels M, Klein J. Adams and Victor's Principles of Neurology. 10th ed. McGraw-Hill Education; 2014.*

Table 1.
Neurodegenerative diseases: main clinical types.

Neurodegeneration represents a loss of neurons and their structural components (dendrites, axons, and synapses) with a corresponding gradual atrophy in neuronal function [2].

NDDs (e.g., Alzheimer's disease, Parkinson's disease, Huntington's disease, dementia with Lewy bodies) determine cognitive and memory deterioration or alteration of the ability to move, speak, and breathe. These chronic and progressive disorders are an important cause of reduced quality of life, morbidity, caregiver burden and also, of the increase in total healthcare expenditure [3–5]. **Table 1** lists the main clinical types of NDDs [6].

Neuroprotection can be defined as a "relative preservation of neuronal structure and/or function" or as an action that aims "to prevent neuronal damage over time (either acute or chronic)" [7]. Neuroprotective action is primary if it is exerted directly on the neuron, or secondary if it appears from an activity on an intermediary that endangers neuronal function [8].

The mechanisms by which most agents with efficiency in NDDs act are not fully elucidated, requiring multiple and in-depth experimental and clinical studies.

Several experimental investigations highlight the multiple and various interrelations between adrenergic, serotoninergic, dopaminergic, glutamatergic, opioid, imidazoline systems, and the nitric oxide pathways, which may elucidate the effects of different compounds involved in the mediation of pathogenic mechanisms responsible for numerous structural, functional, and behavioral disturbances.

This chapter presents a brief overview of the most studied mechanisms related to neuroprotection and details the possibilities to pharmacologically influence through the main known neurotransmitters the pathophysiological mechanisms linked to various NDDs.

2. The imidazoline system

Imidazoline receptors are located not only in the mammalian central nervous system (CNS) cells but also in the peripheral nervous system [9], being involved in the mediation of various physiological processes in the body. It is currently known that there are four types of imidazoline receptors: I1, I2, I3, I4 (non I1-non I2), from which the first three have been mostly studied [10].

It has been emphasized that these receptors play an essential role in cell proliferation, regulation of adipose tissue formation, body temperature maintenance, mediation of gastrointestinal motility, neuroprotection, inflammation, nociceptive sensitivity, and some neurological or psychiatric disorders (such as depression) [11]. Moreover, it is known that these imidazoline receptor subtypes exert control over the activity of the hypothalamic-pituitary-adrenal and noradrenergic axis [12, 13].

A number of different endogenous ligands have been characterized: agmatine, the best known and largely studied, harmane and harmalane (derivatives of the beta-carboline group), and the newly discovered ribotide (acetic acid imidazole). Agmatine, the potent neurotransmitter of the imidazoline system, has an important role in the mediation of body's response to stress, analgesia, drug addiction, and abstinence syndrome, in modulation of seizures development, and in neuroprotection [14, 15].

Endogenous agmatine is produced in response to stress (in conditions of ischemia, prolonged exposure to cold) and/or to inflammation [16]. It is assumed that agmatine is also an effective neurotransmitter, due to its concentration in the brain similar to classical neurotransmitters [17, 18].

Literature data have revealed that agmatine stimulates the activity of endothelial nitric oxide synthase [16], this effect being also proved by its level in the rat brain after cerebral ischemia [19, 20].

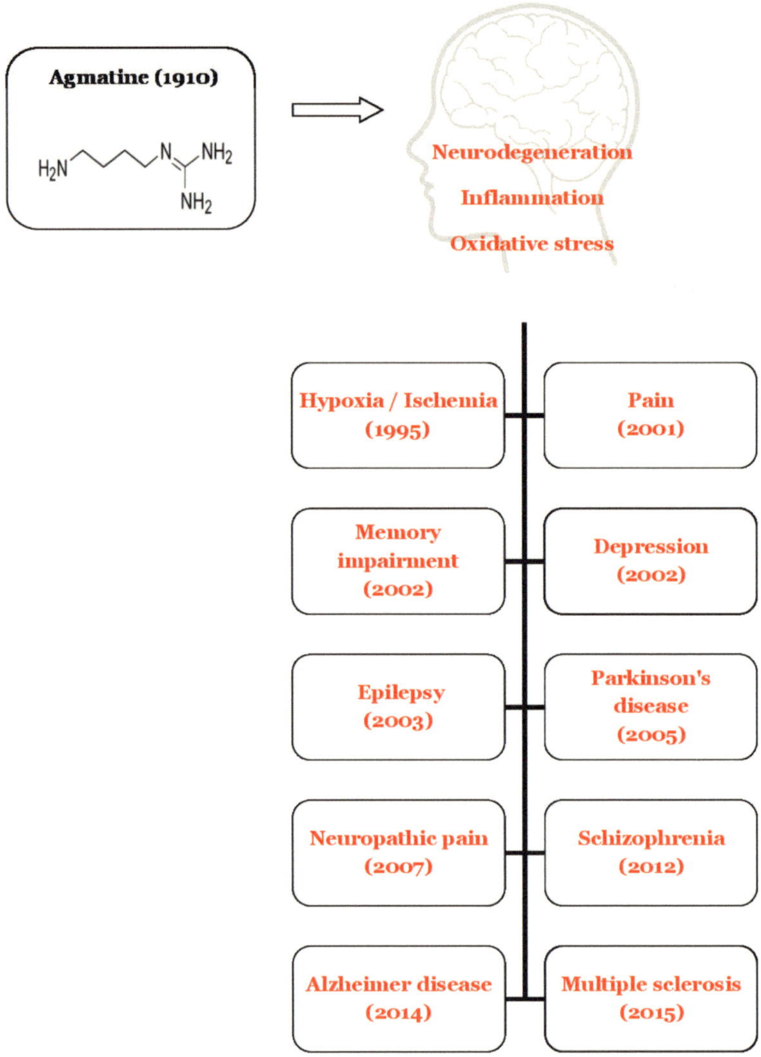

Figure 1.
Neuroprotective effects of agmatine and their first discovery.

Along with evidence of its neuromodulatory and neuroprotective properties, there are numerous preclinical studies demonstrating the beneficial effects of exogenous administration of agmatine in depression, anxiety, hypoxic ischemia, pain, morphine tolerance, memory impairment, Parkinson's disease, Alzheimer's disease, epilepsy, and other related conditions with traumatic brain injuries (**Figure 1**) [21–23]. All these are arguments in favor of the potential of agmatine as a new pharmacological agent for the treatment of various neurological diseases and NDDs [24].

3. The involvement of the imidazoline system in the mediation of cognitive functions

Electrophysiological studies involving various brain areas, performed on laboratory animals with experimentally induced cerebral alterations, have demonstrated the neurotropic effects of agmatine [25].

In vitro experimental researches have shown that activation of I2 receptors via the agmatine endogenous ligand exerts neuroprotective effects by increasing the expression of glial fibrillary acidic protein in astrocyte cultures and by inhibiting MAO activity [26, 27]. Moreover, the beneficial effects of agmatine have been observed on ischemic-hypoxic lesions, on glutamate-induced neurotoxicity by activating the imidazoline receptors [28, 29]. It was also demonstrated that agmatine administration improves learning activity and memory of rats in experimental models of Alzheimer's disease and streptozotocin-induced type two diabetes mellitus [22, 30].

Other experimental investigations highlight the neuroprotective effect of intranasal administration of agmatine in elderly female rats, with a significant improvement of neurological status and increase of survival rate [28, 29].

The neuroprotective effects of agmatine on the morphological changes determined by repeated induced stress on medial prefrontal cortex and hippocampus of the rat were also investigated [31]. It was emphasized that under constant stress conditions, morphological alterations of the brain are associated with the reduction of endogenous agmatine levels [measured by high-performance liquid chromatography (HPLC)] and with an increase of arginine-decarboxylase level in the prefrontal cortex, hippocampus, striatum, and hypothalamus [32].

The exogenous administration of agmatine lowers brain morphological impairment, suggesting thus its neuroprotective effects against structural changes in the rat brain, under recurrent stress circumstances [32]. Moreover, elevated levels of agmatine have been evidenced in the blood, cortex, hippocampus, and hypothalamus, immediately after brain hypoxic ischemia. Other studies emphasize the neuroprotective influences of agmatine, highlighted by the increase of its brain levels, in rats subjected to prolonged cold-exposure stress conditions [33].

The neuroprotective potential of agmatine was also highlighted in the experimental model of 1-methyl-4-phenyltetrahydropyridine (MPTP)-induced Parkinson's disease in mice [29]. The use of agmatine attenuates the loss of cellular dopamine from the black substance and repeated treatment improves short-term memory impairment induced by MPTP in elderly mice. The behavioral benefits of agmatine are associated with the decrease in MPTP-induced glutamate capture in the hippocampal area, suggesting thus its involvement in modulation of glutamate recapture, the possible mechanisms responsible for lowering glutamate extracellular levels, thereby alleviating its neurotoxicity [29].

It is known that alteration of spatial memory in Parkinson's disease and schizophrenia is attributed to several factors, including hypofunction of glutamate and reduction of hippocampal volume [34]. Literature data report that the administration of the N-methyl-D-aspartate (NMDA) receptor antagonists (phencyclidine, also coded MK801) frequently impairs the late alternation performance in a standardized behavioral model of cognitive functions alteration similar to schizophrenia in laboratory animals [34]. The use of the glutamate/NMDA receptor antagonist phencyclidine induces a spectrum of behavioral, neurochemical, and anatomical changes, manifested by locomotor hyperactivity, motor-negative deficits, and cognition alterations (with memory impairment and visual attention) in laboratory animals. This substance was used to induce the experimental schizophrenia in laboratory animals [35, 36]. Agmatine attenuates cognitive and behavioral deficiency in rats with experimental phencyclidine-induced schizophrenic manifestations [37].

The effects of agmatine on memory alterations similar to those found in Alzheimer's disease have been evaluated in rats; in the pathogenesis of this degenerative disorder (which causes cognitive deficits in rodents), the fragment beta amyloid Aβ2 25–35 plays an essential role. Studies have shown that agmatine significantly reduces the alterations in memory and spatial learning induced by the beta amyloid Aβ2 25–35 fragment (the neurotoxic component of beta amyloid Aβ 1–42) in various behavioral experimental models, such as: the swimming test, the radial arm maze test, and the object recognition test [38].

It has been revealed that agmatine diminishes the activation of hippocampal caspase-3 (the early indicator of neuronal apoptosis) and prevents the alteration of spatial memory induced by lipopolysaccharides, in the swimming test in rat [39], suggesting its neuroprotective effects.

The neurotropic activity of agmatine has been also evidenced in the structural and cognitive alterations after the administration of NMDA (N-methyl-D-aspartate) in rats [40]. The use of high performance liquid chromatography (HPLC) and electrochemical detection allowed highlighting that the treatment with NMDA is associated with low concentrations of monoamines (epinephrine, norepinephrine, dopamine, and serotonin) in rat PC12 cells [29, 40]. In this experimental model (swimming test), agmatine protects against NMDA-induced PC12 cell lesions, augmenting the levels of epinephrine, norepinephrine, and dopamine, but not influencing serotonin values, together with lowering intracellular Ca^{2+} overload. These results indicate that the neuroprotective action of agmatine may be related to NMDA-receptor modulation and/or to controlling the decrease in monoamine content and NMDA-induced intracellular Ca^{2+} overload [40].

Immunohistochemical studies and electrophysiological investigations performed on the brain have validated the neuroprotective actions of both imidazoline receptor antagonists idazoxan and efaroxan in rats with cerebral damages caused by the use of quinolinic acid [41], and also in mice with experimentally induced autoimmune encephalomyelitis, confirming the improvement of brain structural alterations and blood brain barrier lesions curtail [42].

A new (+)2-(ethyl-2,3-dihydrobenzofuranyl)-2-imidazoline derivative—dexefaroxan—the (+) enantiomer of efaroxan has been characterized. It has a potent and selective α2 antagonist activity, with facultative effects on cognitive functions in the passive avoidance test in rats with memory-deficiency induced by scopolamine, diazepam, or by the 2-adrenergic agonist UK 14,304.

Dexefaroxan improves the cognitive performances in the passive avoidance test, facilitates spatial memory in the Morris swimming test in rats, and increases the object recognition ability in the specific behavioral test in mice [43, 44]. It has also proved to ameliorate the animal's memory deficits in these tests, particularly by attenuation of spontaneous memory loss, and to improve their spatial recognition ability, rather than through the acquisition skills or various other non-cognitive effects.

After subcutaneous administration of dexefaroxan, its pharmacodynamic effects persist for about 21–25 days, indicating that tolerance does not occur during prolonged treatment. Moreover, it was emphasized that dexefaroxan exerts protective effects on the spatial memory deficit caused by cortical devascularization in the Morris swimming test in rats [43].

Dexefaroxan has also been shown to exhibit neuroprotective effects on the devascularization-induced neurodegeneration, to ameliorate the structural changes in the hippocampus, and to remove the cognitive deficits induced by cerebral ischemia in rats [45, 46]. Its neuroprotective effects were present also in the excitotoxic lesions produced at the region of the basal magnocellular nucleus, increasing the olfactory discriminative capacity of rats, suggesting thus the possibility of its use in the treatment of memory disturbances in Alzheimer's disease [43].

Studies performed on genetically modified animals revealed that dexefaroxan improves cognitive performance in knockout mice with Alzheimer's disease [47].

Literature data regarding the neuroprotective action of imidazoline agonist and antagonist agents in human studies are only few, and the mechanisms involved in these effects are not completely deciphered.

Some investigations suggested that agmatine manifests protective activity against brain cell injury in different *in vivo* models of Parkinson's disease, as well as *in vitro* studies, performed on human-derived dopaminergic neuroblastoma cell lines. It was postulated that the neuromodulatory properties of agmatine are related to the protective effects on the dopaminergic neurons, to NMDA receptor blocking, and to the decrease in oxidative stress, due to the inhibition of nitric oxide synthase (NOS) activity [48, 23].

Other clinical trials highlighted that the treatment with agmatine was associated with cytoprotective actions, in patients with spinal cord injury, proved by lessening the glial weal construction, decreasing the collagen scar zone, relieving the neuronal alterations, and recovering remyelination [49]. Moreover, the beneficial effects of agmatine have been demonstrated in various CNS lesions such as: cerebrovascular accident, brain trauma, neuropathic pain, lumbar degenerative disc disease, and different other types of neuropathy [50–52].

The administration of dexmedetomidine has also shown neuroprotective effects in humans with acute cerebral lesions [53].

In patients with dementia due to brain frontal lesions, idazoxan alleviates attentional and executive dysfunctions evoked by classical cognitive function tests [54].

4. The interrelation between the adrenergic and the imidazoline systems in the mediation of cognitive functions

Clonidine, both a non-specific α2 adrenergic and imidazoline receptor agonist, decreases the cognitive function alterations induced by phencyclidine and MK801, facilitating spatial memory in the radial arm maze test in rats [35], but does not influence the behavioral and cognitive deficits in the experimental NMDA-induced excitotoxic dorsal hippocampal lesions [35]. Such findings indicate that clonidine improves memory alterations caused by glutamate hypofunction, but not by hippocampal injury, implying that multiple and distinct mechanisms are involved in the development of memory disorders.

The administration of the α2 adrenergic receptor agonist clonidine or guanfacine prevents some of the behavioral effects of NMDA antagonists, proving that the monoaminergic system mediates a number of aspects of the cognitive deficit. Clonidine and guanfacine improve the lack of visual attention and spatial memory induced by phencyclidine in rats [34, 36]. It was demonstrated that low doses of clonidine recover the animal's ability to accurately choose the object and prevent the performance deficit induced by phencyclidine. At high doses, clonidine decreases the response time and induces a lack of the choice accuracy. These results indicate that clonidine treatment can alleviate phencyclidine-induced deficit of attention and of working memory, probably by preventing some of the neurochemical and anatomical effects of this psychotomimetic drug [34].

On the other hand, the use of only the selective α2 adrenergic receptor antagonist does not impair the animal's spatial memory, but dramatically aggravates the phencyclidine-induced memory deficit [36]. These data demonstrate that α2 adrenergic receptors mediate the inhibition of spatial memory disturbances, suggesting their important role in cognitive deficits associated to NMDA receptor hypofunction [34, 36].

The role of moxonidine (an α2 adrenergic imidazoline I1 receptor agonist) on cognitive function in rats with Huntington's disease experimentally induced with 3-nitropropionic acid (3-NPA) was investigated in the Morris swimming test and in the elevated plus maze test. The administration of 3-NPA induces degenerative brain damage, progressive motor dysfunction, loss of grip force, emotional disturbances, weight loss, anxiety, and impairment of learning activity and memory. An increase in cerebral acetylcholinesterase level, enhancement of oxidative stress, and impairment of the activity of mitochondrial enzyme complexes I, II, and IV were also noted [28]. The treatment with moxonidine resulted in the alleviation of disturbances caused on animal weight, motor activity, gripping ability, anxiety, impairment of learning ability and memory, and biochemical disturbances, thus indicating that substances modifying the activity of I1 receptors may be potential pharmacological agents for the treatment of degenerative brain disorders [55].

The effects of clonidine have also been evaluated in mice with subacute brain ischemia obtained after permanent ligation of common carotid arteries. The subsequent brain damages consisted of expansion of cerebral infarction areas, assessed by computed tomography scans. This experimentally induced chronic cerebral hypoperfusion was associated with a significant impairment of animal's learning ability and memory in the Morris swimming test [25, 28]. Subacute treatment with clonidine for 7 days increases the expression of neuronal nuclei, glutamic acid-decarboxylase-67, and gamma-aminobutyric acid (GABA) B receptor ($GABA_B1$) in hippocampal subregion cornu amonis (CA_1) but does not influence the level of these elements in the hippocampal area CA_3, nor in the dentate gyrus. These data support the idea that clonidine exerts neuroprotective effects on chronic cerebral ischemic lesions, by regulating $GABA_B1$ receptors and the activity of glutamic acid-decarboxylase-67 [25].

Additionally, the decrease in superoxide dismutase (SOD), catalase (CAT), and glutathione levels as well as the increase of both malondialdehyde (MDA) level and cerebral acetylcholinesterase activity were noted in animals with brain ischemic lesions [28].

Both moxonidine and clonidine have shown a decrease in histopathological changes, oxidative stress, central cholinesterase activity, as well as a reduction in memory disturbances and learning deficits in mice with vascular dementia induced by subacute ischemia after permanent bilateral cerebral artery ligation [28, 40].

In vitro cell culture studies from the rat frontal cortex with glutamate-induced neurotoxicity revealed the partial neuroprotective effects of moxonidine, with a significant decrease in the number of dead cells [26]. Moxonidine has shown beneficial effects on cerebral spasm in an experimental rabbit model of subarachnoid hemorrhage [40, 56].

5. The interrelations between the imidazoline system and the oxidative stress in the mediation of cognitive functions

Different pathological conditions of the body, as well as the physiological process of aging, can cause cognitive impairment and free oxygen radicals production, being responsible for abnormal functioning and cell death. Subsequently, a new idea has emerged claiming that nitric oxide (NO), along with the free radicals, plays a key role in the aging process, due to neurotoxic effects on the brain exerted by its excessive levels [57]. Nitric oxide is generated from L-arginine under the action of nitric oxide synthase (NOS). The three isoforms of NOS have different roles in the body: neuronal NOS (nNOS) is responsible for synaptic plasticity, learning and memory processes; endothelial NOS (eNOS) provides stabilization and regulation of vascular micro-environment and contributes to neuroplasticity [58]; and inducible NOS (iNOS) is involved in various pathophysiological conditions [57].

Numerous experimental researches reveal that NOS activity is significantly elevated in the brain of elderly rats, being associated with existing cognitive alterations [57, 59]. Mediated by the competitive inhibition of nNOS and iNOS, and correlated with the stimulation of NOS, agmatine contributes to the improvement of cognitive functions [19, 60, 61], while exhibiting neuroprotective effects [38, 39].

Literature data have shown that agmatine eliminates neuroinflammation and lipopolysaccharide-induced memory impairment (which is known to stimulate iNOS activity and, implicitly, the NO production) in laboratory animals. It prevents cognitive alterations, probably as a result of inhibition of iNOS activity [39]. Other researchers have disproved these results by showing that agmatine can cause cognitive impairment due to the inhibition of NMDA receptors and of NO, important elements in the modulation of learning and memory processes [62, 63].

On the other hand, it is known that the central cholinergic system plays a crucial role in the mediation of cognitive functions. Cognitive deficits have been induced in laboratory animals by using an anticholinergic agent, scopolamine, its administration producing a significant reduction in NOS activity, and an increase in arginase activity, of L-ornithine and putrescine levels in the hippocampus [51]. It has been observed that agmatine eliminates the scopolamine-induced alterations of memory and learning capacity [64, 65].

Although glutamatergic activity is required for cognitive processes, it is assumed that the increase of glutamate levels or of NMDA activity would also be responsible for the scopolamine-induced cognitive disturbances [66].

Knowing that agmatine blocks NMDA receptors and also interferes with the pathways of NO, NOS, and L-arginine, it was assumed that the removal of scopolamine-induced cognitive deficits can be attributed to its modulating effect on NO/NOS activity, on L-arginine, and also to the antagonization of NMDA receptors, with subsequent suppression of excessive glutamatergic activity [61, 65].

Abnormal release and disturbances of neuromodulatory activities due to variation in cerebral agmatine levels may be correlated to different CNS diseases (such as schizophrenia). Interactions of agmatine with other central neurotransmitter systems (such as glutamatergic and nitrergic) appear to be particularly important in the pathophysiological mechanisms of CNS disorders associated with brain damage and cognitive functions deficit.

6. The opioid system

Neurodegeneration can be caused by chronic disease progression or by acute injury (cerebral ischemia—stroke or trauma) [67]. Ischemic stroke represents a vascular ailment with neurological consequences produced by the obstruction of the arteries in a part of the brain, therefore by blood supply privation [68]. Stroke can determine long-term neurological and psychiatric impairments, its therapy being focused on confining secondary injury processes [67].

In a recent review article, Chamorro presented that ischemic stroke is "the first cause of permanent disability in adult people, the second single most frequent cause of death for people older than 60 years, the second most common cause of dementia, representing approximately 3% to 7% of the total health-care expenditure in high income countries" [69].

A superpose of pathologies in different neurological disorders was proposed since NDDs and ischemic stroke are frequently concomitant, hence the neuroprotective therapy could be similar [70].

Opioids are substances with morphine-like action binding to specific opioid receptors (ORs). In the early 1990s, three important opioid receptor families [μ (MOR), κ (KOR), and δ (DOR)] were identified, and in 1994 another opioid receptor was discovered [nociceptin, orphanin FQ receptor (NOP), or the opioid receptor-like orphan receptor (ORL)]. ORs are found in the nervous system, lungs, heart, liver, and gastrointestinal and reproductive tracts. They have been intensively studied and it was emphasized that they not only are related to

Opioid receptor	Agonist
δ	DADLE [D-Ala2, D-Leu5]-enkephalin
	DPDPE (D-Pen2,D-Pen5)-enkephalin
	SNC80
	Tan-67
	Remifentanil
κ	BRL 52537
	CI-977
	GR89696
	Salvinorin A
	U-50,488H
	Dynorphin
μ	DAMGO [D-Ala2,N-MePhe4,Gly-ol]-enkephalin
	Endorphin 1 and 2 (EM 1/2),
	Morphine

Table 2.
Opioid receptors and their agonists tested for neuroprotective action.

antinociceptive action, but also have a role in cell proliferation, ionic homeostasis, emotional response, immune function, epileptic seizures, feeding, obesity, respiratory and cardiovascular control, hibernation, and neuroprotection [71, 72].

In the last decades, researches have pointed out that the opioid system can be promising to get neuroprotective treatment in the event of stroke, through OR agonists at lower doses, to avoid tolerance and/or physical dependency. DOR agonists followed by KOR agonists have revealed the most intense neuroprotective efficacy [64]. Major OR agonists tested for neuroprotection are listed in **Table 2** [71, 72].

DOR activation is beneficial against ischemic, hypoxic, and excitotoxic injuries [73]; recent studies promote DOR and especially DADLE (an analog of endogenous delta-opioid enkephalin) as promising targets for treating NDDs like stroke and PD [74–76].

DADLE alleviates apoptotic pathways, supports not only cell survival of peripheral organs (such as lung, heart, kidney, and liver) but also neuronal survival, and protects neurons and glial cells from ischemia-induced cell death [76–78].

In a cellular model of PD, DADLE administration augmented cell survivability with concurrent downregulation of the unfolded protein response stress sensors and protein aggregates [79].

Findings from a rat middle cerebral artery occlusion (MCAO) stroke model proposed that neuroprotection of DADLE treatment was based on the activation of PI3K-Akt pathway by reducing nerve cell apoptosis [80].

Non-selective opioid receptor agonists were also tested: LYS739 (fluorinated enkephalin-fentanyl derivative) and the most promising compound—biphalin—which proved to be effective both *in vitro* and *in vivo* stroke models [72].

The latter is a dimeric enkephalin analog (Tyr-D-Ala-Gly-Phe-NH-)2 with high potency and affinity for MOR and DOR and low affinity for KOR. Biphalin crosses blood-brain barrier reaching spinal and supraspinal sites expressing OR and produces less physical dependence and tolerance compared to morphine [72, 81–84].

Different studies using a mouse MCAO stroke model reported that biphalin reduced brain edema and infarction, ROS production, and NMDA-induced excitotoxicity. It also increased locomotor activities and neurological score after stroke resembled to saline-treated animals [72, 81, 85, 86]. Biphalin notably diminished penumbral expression of Na^+, K^+, $2Cl^-$ cotransporter (NKCC), and the translocation of the conventional isoforms of protein kinase C [81].

It has been hypothesized that biphalin's neuroprotective effects are more intense compared to subtype-selective agonists due to concomitant activation of the three types of OR [72, 85].

7. Conclusion

Various pharmacological substances influencing the pathways of the main neurotransmitters have confirmed valuable effects in neurogenesis and neuroprotection, being validated in different *in vitro* researches and *in vivo* experimental animal models of limited or extensive ischemic brain lesions.

Deciphering the roles of the neurotransmitters in central nervous system activity other than the signaling function will represent a starting point to deepen the knowledge about the complex mechanisms of the brain functions and to obtain new agents useful for protection of ischemic neurons and for preventing their irreversible damage.

Author details

Liliana Mititelu-Tartau[1†], Maria Bogdan[2*†], Victor Gheorman[3], Liliana Foia[1], Ancuta Goriuc[1], Gabriela Rusu[1], Beatrice Buca[1], Liliana Pavel[4], Ana Cristofor[1], Cosmin-Gabriel Tartau[1] and Gratiela Eliza Popa[5†]

1 Faculty of Medicine, “Gr. T. Popa” University of Medicine and Pharmacy, Iasi, Romania

2 Faculty of Pharmacy, University of Medicine and Pharmacy, Craiova, Romania

3 Faculty of Medicine, University of Medicine and Pharmacy of Craiova, Romania

4 Faculty of Medicine, “Dunarea de Jos” University of Medicine and Pharmacy, Galati, Romania

5 Faculty of Pharmacy, “Gr. T. Popa” University of Medicine and Pharmacy, Iasi, Romania

*Address all correspondence to: bogdanfmaria81@yahoo.com

† These authors contributed equally to this work.

References

[1] Jaafaru MS, Abd Karim NA, Enas ME, Rollin P, Mazzon E, Abdull Razis AF. Protective effect of glucosinolates hydrolytic products in neurodegenerative diseases (NDDs). Nutrients. 2018;**10**(5), 580:1-15

[2] Cummings J. Disease modification and neuroprotection in neurodegenerative disorders. Translational Neurodegeneration. 2017;**6**:25

[3] Lauterbach EC, Victoroff J, Coburn KL, Shillcutt SD, Doonan SM, Mendez MF. Psychopharmacological neuroprotection in neurodegenerative disease: Assessing the preclinical data. The Journal of Neuropsychiatry and Clinical Neurosciences. 2010;**22**:8-18

[4] Gitler AD, Dhillon P, Shorter J. Neurodegenerative disease: Models, mechanisms, and a new hope. Disease Models & Mechanisms. 2017;**10**:499-502

[5] Gao HM, Hong JS. Why neurodegenerative diseases are progressive: Uncontrolled inflammation drives disease progression. Trends in Immunology. 2008;**29**:357-365

[6] Ropper A, Samuels M, Klein J. Adams and Victor's Principles of Neurology. 10th ed. Boston: McGraw-Hill Education; 2014. pp. 1060-1131

[7] Wiendl H, Elger C, Förstl H, Hartung H-P, Oertel W, Reichmann H, et al. Gaps between aims and achievements in therapeutic modification of neuronal damage ("Neuroprotection"). Neurotherapeutics. 2015;**12**:449-454

[8] Cummings J, Fox N. Defining disease modifying therapy for Alzheimer's disease. The Journal of Prevention of Alzheimer's Disease. 2017;**4**:109-115

[9] Ferrari F, Fiorentino S, Mennuni L, Garofalo P, Letari O, Mandelli S, et al. Analgesic efficacy of CR4056, a novel imidazoline-2 receptor ligand, in rat models of inflammatory and neuropathic pain. Journal of Pain Research. 2011;**4**:111-125

[10] Nechifor M. Imidazoline receptors—Normal and pathological factors. Revista Medico-Chirurgicala a Societatii de Medici Si Naturalisti din Iasi. 2001;**105**: 438-443

[11] Dardonville C, Rozas I, Callado LF, Meana JJ. I(2)-imidazoline binding site affinity of a structurally different type of ligands. Bioorganic & Medicinal Chemistry. 2002;**10**:1525-1533

[12] Pypendop BH, Silverstein D, Hopper K. 2 agonists and antagonists. In: Small Animal Critical Care Medicine. 2nd ed. Missouri: Saunders Inc.; 2015. pp. 866-871.Ch 165. SRC - BaiduScholar

[13] Thorn DA, An X-F, Zhang Y, Pigini M, Li J-X. Characterization of the hypothermic effects of imidazoline I2 receptor agonists in rats. British Journal of Pharmacology. 2012;**166**:1936-1945

[14] Smith KL, Jessop DS, Finn DP. Modulation of stress by imidazoline binding sites: Implications for psychiatric disorders. Stress (Amsterdam, Netherlands). 2009;**12**:97-114

[15] Ciubotariu D, Nechifor M. Involvement of imidazoline system in drug addiction. Revista Medico-Chirurgicala a Societatii de Medici Si Naturalisti din Iasi. 2012;**116**:1118-1122

[16] Halaris A, Plietz J. Agmatine: Metabolic pathway and spectrum of activity in brain. CNS Drugs. 2007;**21**: 885-900

[17] Haenisch B, von Kügelgen I, Bönisch H, Göthert M, Sauerbruch T, Schepke M, et al. Regulatory mechanisms underlying agmatine homeostasis in

humans. American Journal of Physiology. Gastrointestinal and Liver Physiology. 2008;**295**:G1104-G1110

[18] Reis DJ, Regunathan S. Is agmatine a novel neurotransmitter in brain? Trends in Pharmacological Sciences. 2000;**21**: 187-193

[19] Mun CH, Lee WT, Park KA, Lee JE. Regulation of endothelial nitric oxide synthase by agmatine after transient global cerebral ischemia in rat brain. Anatomy & Cell Biology. 2010;**43**: 230-240

[20] Nissim I, Horyn O, Daikhin Y, Chen P, Li C, Wehrli SL, et al. The molecular and metabolic influence of long term agmatine consumption. The Journal of Biological Chemistry. 2014;**289**: 9710-9729

[21] Mancinelli F, Ragonese F, Cataldi S, Ceccarini MR, Iannitti RG, Arcuri C, et al. Inhibitory effects of agmatine on monoamine oxidase (MAO) activity: Reconciling the discrepancies. The EuroBiotech Journal. 2018;**2**:121-127

[22] Sirvanci-Yalabik M, Sehirli AO, Utkan T, Aricioglu F. Agmatine, a metabolite of arginine, improves learning and memory in streptozotocin-induced Alzheimer's disease model in rats. Bulletin of Clinical Psychopharmacology. 2016;**26**(4): 342-354

[23] Neis VB, Rosa PB, Olescowicz G, Rodrigues ALS. Therapeutic potential of agmatine for CNS disorders. Neurochemistry International. 2017; **108**:318e331

[24] Moretti M, Neis VB, Matheus FC, Cunha MP, Rosa PB, Ribeiro CM, et al. Effects of agmatine on depressive-like behavior induced by intracerebroventricular administration of 1-methyl-4-phenylpyridinium (MPP (+)). Neurotoxicity Research. 2015;**28**: 222-231

[25] Lu Y, Li C, Zhou M, Luo P, Huang P, Tan J, et al. Clonidine ameliorates cognitive impairment induced by chronic cerebral hypoperfusion via up-regulation of the $GABA_BR1$ and GAD67 in hippocampal CA1 in rats. Pharmacology, Biochemistry, and Behavior. 2015;**132**:96-102

[26] Bakuridze K, Savli E, Gongadze N, Baş DB, Gepdiremen A. Protection in glutamate-induced neurotoxicity by imidazoline receptor agonist moxonidine. The International Journal of Neuroscience. 2009;**119**:1705-1717

[27] Head GA, Mayorov DN. Imidazoline receptors, novel agents and therapeutic potential. Cardiovascular & Hematological Agents in Medicinal Chemistry. 2006;**4**:17-32

[28] Gupta S, Sharma B. Pharmacological modulation of I(1)-imidazoline and α2-adrenoceptors in sub acute brain ischemia induced vascular dementia. European Journal of Pharmacology. 2014;**723**:80-90

[29] Matheus FC, Aguiar AS, Castro AA, Villarinho JG, Ferreira J, Figueiredo CP, et al. Neuroprotective effects of agmatine in mice infused with a single intranasal administration of 1-methyl-4-phenyl-1,2,3,6-tetrahydropyridine (MPTP). Behavioural Brain Research. 2012;**235**:263-272

[30] Kang S, Kim C-H, Jung H, Kim E, Song H-T, Lee JE. Agmatine ameliorates typ. 2 diabetes induced-Alzheimer's disease-like alterations in high-fat diet-fed mice via reactivation of blunted insulin signalling. Neuropharmacology. 2017;**113**:467-479

[31] Aricioglu F, Ercil E, Dulger G. Agmatine inhibits naloxone-induced contractions in morphine-dependent Guinea pig ileum. Annals of the New York Academy of Sciences. 2003;**1009**: 147-151

[32] Zhu M-Y, Wang W-P, Cai Z-W, Regunathan S, Ordway G. Exogenous agmatine has neuroprotective effects against restraint-induced structural changes in the rat brain. The European Journal of Neuroscience. 2008;**27**: 1320-1332

[33] Aricioglu F, Regunathan S. Agmatine attenuates stress- and lipopolysaccharide-induced fever in rats. Physiology & Behavior. 2005;**85**: 370-375

[34] Jeutsch JD, Anzivino LA. A low dose of the alpha2 agonist clonidine ameliorates the visual attention and spatial working memory deficits produced by phencyclidine administration to rats. Psychopharmacology (Berl). 2004;**175**: 76-83

[35] Bardget ME, Points M, Ramsey-Faulkner C. The effects of clonidine on discrete-trial delayed spatial alternation in two rat models of memory loss. Neuropsychopharmacology. 2008;**33**: 1980-1991

[36] Marrs W, Kuperman J, Avedian T, Roth RH, Jentsch JD. Alpha-2 adrenoceptor activation inhibits phencyclidine-induced deficits of spatial working memory in rats. Neuropsychopharmacology. 2005;**30**: 1500-1510

[37] Unal G, Ates A, Aricioglu F. Agmatine-attenuated cognitive and social deficits in subchronic MK-801 model of schizophrenia in rats. Psychiatry and Clinical Psychopharmacology. 2018:1-10

[38] Bergin DH, Liu P. Agmatine protects against beta-amyloid25–35-induced memory impairments in the rat. Neuroscience. 2010;**169**:794-811

[39] Zarifkar A, Choopani S, Ghasemi R, Naghdi N, Maghsoudi AH, Maghsoudi N, et al. Agmatine prevents LPS-induced spatial memory impairment and hippocampal apoptosis. European Journal of Pharmacology. 2010;**634**: 84-88

[40] Li YF, Gong ZH, Cao JB, Wang HL, Luo ZP, Li J. Antidepressant-like effect of agmatine and its possible mechanism. European Journal of Pharmacology. 2003;**469**(1–3):81-88

[41] Martel J, Chopin P, Colpaert F, Marien M. Neuroprotective effects of the alpha2-adrenoceptor antagonists, (+)-efaroxan and (+/−)-idazoxan, against quinolinic acid-induced lesions of the rat striatum. Experimental Neurology. 1998;**154**:595-601

[42] Wang X-S, Fang H-L, Chen Y, Liang S-S, Zhu Z-G, Zeng Q-Y, et al. Idazoxan reduces blood-brain barrier damage during experimental autoimmune encephalomyelitis in mouse. European Journal of Pharmacology. 2014;**736**: 70-76

[43] Chopin P, Debeir T, Raisman-Vozari R, Colpaert FC, Marien MR. Protective effect of the alpha2-adrenoceptor antagonist, dexefaroxan, against spatial memory deficit induced by cortical devascularization in the adult rat. Experimental Neurology. 2004;**185**: 198-200

[44] Pauwels PJ, Rauly I, Wurch T. Dissimilar pharmacological responses by a new series of imidazoline derivatives at precoupled and ligand-activated alpha 2A-adrenoceptor states: Evidence for effector pathway-dependent differential antagonism. The Journal of Pharmacology and Experimental Therapeutics. 2003;**305**:1015-1023

[45] Rizk P, Salazar J, Raisman-Vozari R, Marien M, Ruberg M, Colpaert F, et al. The alpha2-adrenoceptor antagonist dexefaroxan enhances hippocampal neurogenesis by increasing the survival and differentiation of new granule cells. Neuropsychopharmacology. 2006;**31**: 1146-1157

[46] Debeir T, Marien M, Ferrario J, Rizk P, Prigent A, Colpaert F, et al. In vivo upregulation of endogenous NGF in the rat brain by the alpha2-adrenoreceptor antagonist dexefaroxan: Potential role in the protection of the basalocortical cholinergic system during neurodegeneration. Experimental Neurology. 2004;**190**:384-395

[47] Francis BM, Yang J, Hajderi E, Brown ME, Michalski B, McLaurin J, et al. Reduced tissue levels of noradrenaline are associated with behavioral phenotypes of the TgCRND8 mouse model of Alzheimer's disease. Neuropsychopharmacology. 2012;**37**: 1934-1944

[48] Condello S, Monica Currò M, Ferlazzo N, Caccamo D, Satriano J, Lentile R. Agmatine effects on mitochondrial membrane potential and NF-κB activation protect against rotenone-induced cell damage in human neuronal-like SH-SY5Y cells. Journal of Neurochemistry. 2011;**116**(1):67-75

[49] Park YM, Lee WT, Bokara KK, Seo SK, Park SH, Kim JH, et al. The multifaceted effects of agmatine on functional recovery after spinal cord injury through modulations of BMP-2/4/7 expressions in neurons and glial cells. PLoS One. 2013;**8**:e53911

[50] Tohidi V, Hassanzadeh B, Sherwood B, Ma W, Rosenberg M, Gilad V, et al. Effect of agmatine sulfate on neuropathic pain. Neurology. 2014;**82** (Suppl. 10):P7.094

[51] Piletz JE, Aricioglu F, Cheng JT, Fairbanks CA, Gilad VH, Haenisch B, et al. Agmatine: Clinical applications after 100 years in translation. Drug Discovery Today. 2013;**18**(17–18): 880-893

[52] Keynan O, Mirovsky Y, Dekel S, Gilad VH, Gilad GM. Safety and efficacy of dietary agmatine sulfate in lumbar disc-associated radiculopathy. An open-label, dose-escalating study followed by a randomized, double-blind, placebo-controlled trial. Pain Medicine. 2010; **11**(3):356-368

[53] Himmelseher S, Kochs EF. Neuroprotection by dexmedetomidine. In: Gullo A, editor. Anaesthesia, Pain, Intensive Care and Emergency A.P.I.C.E. Italia: Springer-Verlag; 2006. pp. 633-639. Ch. 56

[54] Coull JT, Sahakian BJ, Hodges JR. The α2 antagonist idazoxan remediates certain attentional and executive dysfunction in patients with dementia of frontal type. Psychopharmacology. 1996;**123**:239-249

[55] Gupta S, Sharma B. Pharmacological benefit of I(1)-imidazoline receptors activation and nuclear factor kappa-B (NF-κB) modulation in experimental Huntington's disease. Brain Research Bulletin. 2014;**102**:57-68

[56] Ilik MK, Kocaogullar Y, Koc O, Esen H. Beneficial effects of moxonidine on cerebral vasospasm after experimental subarachnoid. Turkish Neurosurgery. 2014;**24**(6):873-879

[57] Liu P, Smith PF, Appleton I, Darlington CL, Bilkey DK. Potential involvement of NOS and arginase in age-related behavioural impairments. Experimental Gerontology. 2004;**39**: 1207-1222

[58] Rushaidhi M, Collie ND, Zhang H, Liu P. Agmatine selectively improves behavioural function in aged male Sprague-Dawley rats. Neuroscience. 2012;**218**:206-215

[59] Gupta N, Jing Y, Collie ND, Zhang H, Liu P. Ageing alters behavioural function and brain arginine metabolism in male Sprague-Dawley rats. Neuroscience. 2012;**226**:178-196

[60] Liu P, Jing Y, Collie ND, Chary S, Zhang H. Memory-related changes in

L-citrulline and agmatine in the rat brain. Hippocampus. 2009;**19**:597-602

[61] Moosavi M, Khales GY, Abbasi L, Zarifkar A, Rastegar K. Agmatine protects against scopolamine-induced water maze performance impairment and hippocampal ERK and Akt inactivation. Neuropharmacology. 2012;**62**(5–6):2018-2023

[62] Stewart LS, McKay BE. Acquisition deficit and time-dependent retrograde amnesia for contextual fear conditioning in agmatine-treated rats. Behavioural Pharmacology. 2000;**11**: 93-97

[63] McKay BE, Lado WE, Martin LJ, Galic MA, Fournier NM. Learning and memory in agmatine-treated rats. Pharmacology, Biochemistry, and Behavior. 2002;**72**:551-557

[64] Utkan T, Gocmez SS, Regunathan S, Aricioglu F. Agmatine, a metabolite of L-arginine, reverses scopolamine-induced learning and memory impairment in rats. Pharmacology, Biochemistry, and Behavior. 2012;**102**: 578-584

[65] Knox LT, Jing Y, Fleete MS, Collie ND, Zhang H, Liu P. Scopolamine impairs behavioural function and arginine metabolism in the rat dentate gyrus. Neuropharmacology. 2011;**61**: 1452-1462

[66] Barber TA, Haggarty MK. Memantine ameliorates scopolamine-induced amnesia in chicks trained on taste-avoidance learning. Neurobiology of Learning and Memory. 2010;**93**: 540-545

[67] Nguyen L, Lucke-Wold BP, Mookerjee S, Kaushal N, Matsumoto RR. Sigma-1 receptors and neurodegenerative diseases: Towards a hypothesis of sigma-1 receptors as amplifiers of neurodegeneration and neuroprotection. Advances in Experimental Medicine and Biology. 2017;**964**:133-152

[68] Xiong XY, Liu L, Yang QW. Refocusing neuroprotection in cerebral reperfusion era: New challenges and strategies. Frontiers in Neurology. 2018; **9**:249

[69] Chamorro A. Neuroprotectants in the era of reperfusion therapy. Journal of Stroke. 2018;**20**(2):197-207

[70] Wu Q J, Tymianski M. Targeting NMDA receptors in stroke: New hope in neuroprotection. Molecular Brain. 2018; **11**(1):15

[71] Feng Y, He X, Yang Y, Chao D, Lazarus LH, Xia Y. Current research on opioid receptor function. Current Drug Targets. 2012;**13**:230-246

[72] Vaidya B, Sifat AE, Karamyan VT, Abbruscato TJ. The neuroprotective role of the brain opioid system in stroke injury. Drug Discovery Today. 2018;**23**: 1385-1395

[73] Yang Y, Zhi F, He X, Moore ML, Kang X, Chao D, et al. Delta-opioid receptor activation and microRNA expression of the rat cortex in hypoxia. PLoS One. 2012;7:e51524

[74] Grant Liska M, Crowley MG, Lippert T, Corey S, Borlongan CV. Delta opioid receptor and peptide: A dynamic therapy for stroke and other neurological disorders. In: Handbook of Experimental Pharmacology. Berlin, Heidelberg: Springer; 2017

[75] Staples M, Acosta S, Tajiri N, Pabon M, Kaneko Y, Borlongan CV. Delta opioid receptor and its peptide: A receptor-ligand neuroprotection. International Journal of Molecular Sciences. 2013;**14**(9):17410-17419

[76] Liu H, Chen B, Li S, Yao J. Dose-dependent neuroprotection of delta-opioid peptide [D-Ala(2), D-Leu(5)]

enkephalin on spinal cord ischemia-reperfusion injury by regional perfusion into the abdominal aorta in rabbits. Journal of Vascular Surgery. 2016;**63**(4): 1074-1081

[77] Lee JY, Liska MG, Crowley M, Xu K, Acosta SA, Borlongan CV, et al. Multifaceted effects of delta opioid receptors and DADLE in diseases of the nervous system. Current Drug Discovery Technologies. 2018;**15**(2): 94-108

[78] Crowley MG, Liska MG, Lippert T, Corey S, Borlongan CV. Utilizing delta opioid receptors and peptides for cytoprotection: Implications in stroke and other neurological disorders. CNS & Neurological Disorders Drug Targets. 2017;**16**(4):414-424

[79] Moghal ETB, Venkatesh K, Sen D. The delta opioid peptide D-Alanine 2, Leucine 5 Enkephaline (DADLE)-induces neuroprotection through cross-talk between the UPR and pro-survival MAPK-NGF-Bcl2 signaling pathways via modulation of several micro-RNAs in SH-SY5Y cells subjected to ER stress. Cell Biology International. 2018;**42**(5): 543-569

[80] Lv MR, Li B, Wang MG, Meng FG, Yu JJ, Guo F, et al. Activation of the PI3K-Akt pathway promotes neuroprotection of the delta-opioid receptor agonist against cerebral ischemia-reperfusion injury in rat models. Biomedicine & Pharmacotherapy. 2017;**93**:230-237

[81] Yang L, Shah K, Wang H, Karamyan VT, Abbruscato TJ. Characterization of neuroprotective effects of biphalin, an opioid receptor agonist, in a model of focal brain ischemia. The Journal of Pharmacology and Experimental Therapeutics. 2011;**339**(2):499-508

[82] Kawalec M, Kowalczyk JE, Beresewicz M, Lipkowski AW, Zablocka B. Neuroprotective potential of biphalin, multireceptor opioid peptide, against excitotoxic injury in hippocampal organotypic culture. Neurochemical Research. 2011;**36**(11): 2091-2095

[83] Remesic M, Macedonio G, Mollica A, Porreca F, Hruby V, Lee YS. Cyclic biphalin analogues with a novel linker lead to potent agonist activities at mu, delta, and kappa opioid receptors. Bioorganic & Medicinal Chemistry. 2018;**26**(12):3664-3667

[84] Popiolek-Barczyk K, Piotrowska A, Makuch W, Mika J. Biphalin, a dimeric enkephalin, alleviates LPS-induced activation in rat primary microglial cultures in opioid receptor-dependent and receptor-independent manners. Neural Plasticity. 2017;**2017**:3829472

[85] Yang L, Islam MR, Karamyan VT, Abbruscato TJ. In vitro and in vivo efficacy of a potent opioid receptor agonist, biphalin, compared to subtype-selective opioid receptor agonists for stroke treatment. Brain Research. 1609; **2015**:1-11

[86] Islam MR, Yang L, Lee YS, Hruby VJ, Karamyan VT, Abbruscato TJ. Enkephalin-fentanyl multifunctional opioids as potential neuroprotectants for ischemic stroke treatment. Current Pharmaceutical Design. 2016;**22**: 6459-6468

Chapter 3

The Role and Development of the Antagonist of Adenosine A_{2A} in Parkinson's Disease

Widya Dwi Aryati, Nabilah Nurtika Salamah, Rezi Riadhi Syahdi and Arry Yanuar

Abstract

Adenosine is a neuromodulator that regulates the body's response to dopamine and another neurotransmitter in the brain that is responsible for motoric, emotion, learning, and memory function. Adenosine is a G-protein-coupled receptor and has four subtypes, which are A_1, A_{2A}, A_{2B}, and A_3. Adenosine A_{2A} is located in the striatum of the brain. Antagonist interferes with GABA releasing, modulates acetylcholine and releases dopamine, and also facilitates dopamine receptor's signaling. Therefore, it can reduce motoric symptoms in Parkinson's disease. Adenosine A_{2A} antagonist is also believed to have neuroprotective effects. Several compounds have been reported and have undergone clinical test as selective adenosine A_{2A} antagonists, including istradefylline, preladenant, tozadenant, vipadenant, ST-1535, and SYN-115. Nonselective adenosine A_{2A} antagonists from natural compounds are caffeine and theophylline.

Keywords: adenosine A_{2A}, selective adenosine A_{2A} antagonists, Parkinson's disease, neuroprotective, natural compounds

1. Introduction

Adenosine is a neuromodulator that coordinates responses to dopamine and other neurotransmitters in areas of the brain responsible for motor function, mood, learning, and memory [1]. Adenosine consists of four receptor subtypes: A_1, A_{2A}, A_{2B}, and A_3 belonging to the superfamily of G-protein-coupled receptor. Adenosine A_1 and A_3 receptors are coupled to inhibitory G proteins, while A_{2A} and A_{2B} receptors are coupled to stimulatory G proteins [2].

Adenosine A_1 receptor can be found in adipose tissue, heart muscle, and inflammatory cells. The receptor mostly expressed by the central nervous system such as neocortex, cerebellum, hippocampus, and dorsal horn of the spinal cord [3]. The pre- and postsynaptic nerve terminals, mast cells, airway smooth muscle, and circulating leukocytes are the places where adenosine A_2 receptor can be found. As the more widely dispersed receptor, adenosine A_2 is divided into two receptors on the basis of high- and low-affinity for adenosine, A_{2A} and A_{2B} [4]. Striatal neurons are where the adenosine A_{2A} are highly enriched; however its lower levels can also be found in glial cells and neurons outside the striatum [5]. The adenosine A_{2B} receptors are highly expressed in the gastrointestinal tract, bladder, lung, and on mast cells. The most widely dispersed receptor is the A_3 receptor which can be found in the kidney, testis, lung, mast cells, eosinophils, neutrophils, heart, and the brain cortex [4].

Adenosine A_{2A} receptors are found to be concentrates in GABAergic medium-sized spiny neurons in the dopamine-rich regions of the brain. The protein translated in the adenosine A_{2A} is carried by many other tissues such as blood vessels, endothelial, lymphoid cells, smooth muscle cells, and several neurons in sympathetic and parasympathetic systems [6]. Therefore, the dispersion of adenosine A_{2A} is not limited to the medium spiny neurons in the basal ganglia. It stimulates the modulation of cAMP production and increases the level of adenylyl cyclase. This receptor is essential in giving the medium of vasodilation of coronary arteries which then supports the combination of new blood vessels and giving protection for tissues from indirect inflammatory damage [7]. The role of the A_{2A} in the brain includes influencing the activity within the indirect pathway of the basal ganglia. The A_{2A} has complicated actions because it colocalizes and is physically combined with other unrelated G-protein-coupled receptors. Therefore, it can form heterodimers such as dopamine D_2/A_{2A}, and D_3/A_{2A}, cannabinoid CB_1/A_{2A}, and glutamate $mGluR5/A_{2A}$, as well as $CB_1/A_{2A}/D_2$ heterotrimers [7].

The pathways which give signals used by the A_{2A} receptor depend on the location of the cell and tissue, the specific G protein which couples it, and the signaling in the cell. The brain also carries the A_{2A} receptor in which it plays an important role in regulating the glutamate and releasing the dopamine [8]. In the striatopallidal neurons, dopamine D_2 receptors are colocalized with adenosine A_{2A} receptors. Adenosine A_{2A} receptor activity that mediates stimulation and D_2 receptors that mediate inhibition in the striatopallidal pathway are balanced [9]. The adenosine A_{2A}

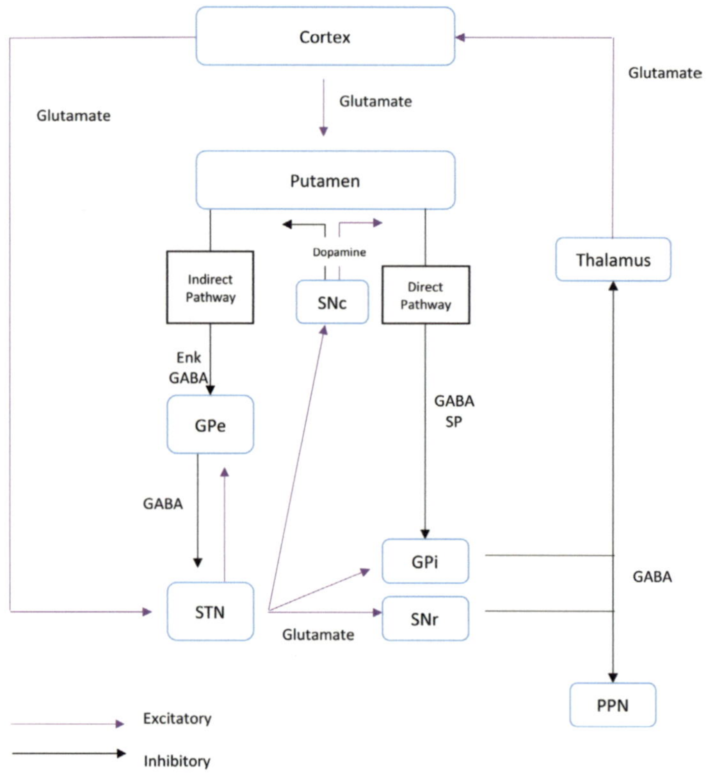

Figure 1.
Basal ganglia circuitry in normal conditions.

likely affects motor activity by acting at different levels of the basal ganglia network. The basal ganglia comprise the striatum (putamen), the globus pallidus externa (GPe), the globus pallidus interna (GPi), substantia nigra pars compacta (SNc), substantia nigra reticulata (SNr), and the subthalamic nucleus (STN). The striatum is represented by medium-sized spiny projection neurons (MSNs), accounting for almost 95% of striatal neurons and using γ-aminobutyric acid (GABA) as neurotransmitter. The GABAergic spiny neurons give rise to the two main striatal efferent circuits: the striatonigral and the striatopallidal pathway. The neurons of the striatonigral (direct) pathway contain the neuropeptide substance P and dynorphin and mainly express D_1 receptors; this pathway directly projects from the striatum to the GPi/SNr. The neurons of the striatopallidal (indirect) pathway containing the neuropeptide, enkephalin (ENK), predominantly express D_2 receptors; this circuit connects the striatum with the GPi/SNr via synaptic connections in the GPe and STN in **Figure 1**. Dopamine modulates motor coordination and fine movements by facilitating the action of the direct pathway on stimulatory D_1 receptors and by inhibiting indirect pathway function acting on inhibitory D_2 receptors [10].

The adenosine A_{2A} receptor has agonists and antagonists of which the roles are potentiating and inhibiting, respectively. The D_2 receptor agonist has effects on motor activity, the releasing of neurotransmitter, and the expression of striatal of c-Fos, a factor of transcription which is used as neuronal activity's indirect marker [11]. The adenosine A_{2A} receptor has a key role in regulating the striatal dopaminergic neurotransmission which produces substances that are valuable to treat neurological disorders that are relevant with dopaminergic dysfunction.

The topology of G-protein-coupled receptor is displayed in the structure of the adenosine A_{2A} receptor. These receptors have a central core which consists of seven

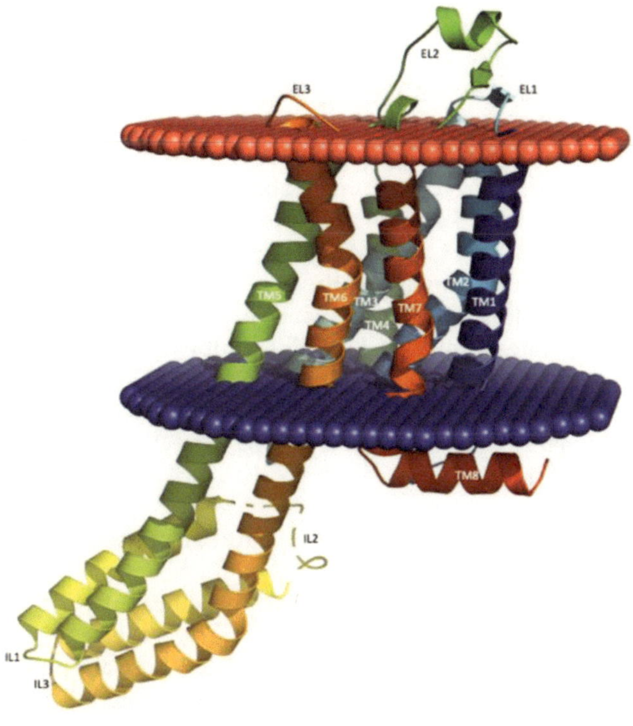

Figure 2.
Crystal structure of the adenosine A_{2A} receptor (4EIY) shown in the membrane structure. The extracellular and intracellular parts of the membrane are shown in red and blue beads, respectively. The disorder residues of intracellular loop (IL2) are modeled in dashed line.

transmembrane helices (7TM). Each of the TM is mainly α-helical and consists of 20–27 amino acids. Three intracellular (IL1, IL2, and IL3) and three extracellular (EL1, EL2, and EL3) loops connect each of the TM domain. A short helix TM8 runs parallel to the cytoplasmic surface of the membrane. The adenosine A_{2A} receptor has differences in length and N-terminal extracellular domain function, their domain of C-terminal intracellular, and their loops of intracellular/extracellular. These differences are shown in **Figure 2**.

2. The role of adenosine A_{2A} in Parkinson's disease

Parkinson's disease (PD) is a chronic neurodegenerative disorder in the brain, marked by motoric symptoms [12]. The motoric symptoms in PD are resting tremor, rigidity, bradykinesia, and postural disorder. Besides motoric symptoms, PD also has non-motoric symptoms such as depression, hallucination, sleeping disorder, and decreasing cognitive and sensory functions. The main pathological characteristic of PD is the loss of dopaminergic neurons in *substantia nigra pars compacta*, a region in the brain that controls all the body movement and forms the dopamine. The development of PD also includes the formation of Lewy body, a deposit of cytoplasmic, eosinophilic neuronal inclusions, composed of the presynaptic protein α-synuclein [13, 14].

The current therapy of PD is targeted at dopamine replacement, thereby decreasing the motor symptoms. It includes precursor of dopamine (levodopa), dopamine agonists [15, 16] monoamine oxidase type B (MAO-B) inhibitors [17], and catechol-O-methyltransferase (COMT) inhibitors [17, 18]. These agents produce undesirable side effects such as on-off effects, hallucinations, and dyskinesia. These effects get more severe as the treatment continued. The efficacy of these agents is also decreasing as the disease progressed [19].

Because of the undesirable side effects of dopamine replacement therapy, the non-dopaminergic therapy is continuously being explored. One of the approaches is selective adenosine A_{2A} antagonist [20, 21]. Adenosine A_{2A} receptors are found mainly in the striatum of rat [22, 23], which has similar distribution with the human brain [24, 25]. In the striatum, adenosine A_{2A} receptors are colocalized with dopamine D_2 receptors. These two receptors have opposite effect on motoric function [26]. The activation of adenosine A_{2A} receptors will inhibit the signaling of dopamine D_2 receptors, and conversely, the inhibition of signaling of adenosine A_{2A} receptors will increase the activation of dopamine D_2 receptors, therefore facilitating dopamine D_2-mediated responses [11]. The inhibition of adenosine A_{2A} receptors showed motoric improvement in animal models of PD [27–30]. This also has desirable effect on long-term levodopa treatment such as decreasing the dyskinesia and increasing the therapeutic effect on levodopa [31, 32].

3. Adenosine A_{2A} receptor antagonist as a neuroprotective

For years, adenosine-dopamine interactions have been investigated in order to observe their relevance for treatment of central nervous system (CNS) disorders [33]. It is assumed that adenosine A_1 receptors (A_1Rs) play an important role in neuroprotection as their activation at the onset of neuronal injury has shown to reduce brain damage in adult animal model. Vice versa, their blockade aggravates the damage. In other hand, adenosine A_2 receptors ($A_{2A}Rs$) are shown to be upregulated in harmful brain conditions, and their blockade shows brain neuroprotection in studied animals [34]. The blockade of $A_{2A}Rs$ alleviates the long-term burden of brain disorders in different neurodegenerative conditions, namely, ischemia, epilepsy, and Parkinson's and Alzheimer's disease, through its control on neuronal cell death [35].

A_{2A}Rs have been shown to be viable in serving as alternative non-dopaminergic strategy of Parkinson's disease treatment because of their limited distribution in the striatum and the intense interaction between adenosine and dopamine receptors in the brain. A_{2A}Rs antagonists were shown to improve motor function in different animal models (primates and rodents), alone or co-administered with dopaminomimetic drugs, levodopa, or dopamine agonists [35]. Based on rigorous preclinical animal studies, istradefylline (KW6002) has shown its promising ability to increase motor activity in PD of the advanced stage in clinical phase IIB trial [36]. It became the first therapeutic agent developed to target A_{2A}Rs, and other similar compounds will be available in near future [37].

The recent meta-analysis (n = 6) suggested that 20 mg of istradefylline improves unified Parkinson's disease ranking scale (UPDRS) III. Meanwhile at 40 mg per day, istradefylline could alleviate off time and motor symptoms derived from Parkinson's disease [38]. Phase 3 study (613 randomized patients), done by Isaacson et al. concluded that greater reduction from baseline in total hours off time/day were shown at all-time points for istradefylline 20 and 40 mg/day, compared to placebo. However, future development is needed as the study has not yet reached statistical significance [39].

In the case of Parkinson's disease, microglia has been suggested to be the most likely cell type to be targeted by A_{2A}Rs antagonists [40]. In vitro and in vivo studies showed that local neuroinflammation make glial cells (especially microglial cells) particularly sensitive to A_{2A}R modulation [41]. Previous research done by Gao and Phillis is the first study to demonstrate nonselective A_{2A}R antagonist action in reducing cerebral ischemic injury in the gerbil, following global forebrain ischemia [42]. After that, many studies have reported the neuroprotective of A_{2A}R antagonists in different models of ischemia [43].

Alzheimer's disease (AD) is a chronic neurodegenerative disorder that is indicated by the progressive loss of memory and other cognitive functions, leading to dementia [44, 45]. Adenosine can control and integrate cognition and memory [46]. Both A_1Rs and A_{2A}Rs, mainly located in synapses, control the release of neurotransmitters which are involved in memory or other cognitive processes [34, 47]. Methylxanthine was discovered to act as nonselective adenosine receptors antagonist. Caffeine, the most famous methylxanthine found in common beverages, is the most widely consumed psychoactive drug. Maia and de Mendonca presented the first epidemiological data showing that the incidence of AD is inversely proportioned with coffee consumption [48]. Several other studies also show this inverse relationship [49–51]. Animal models also shown that caffeine intake may be beneficial for AD. In a study, a 6-month period of 0.3 g/L caffeine intake alleviated the cognitive deficits found in AD transgenic mice (APPsw). Furthermore, these mice culture neurons showed the reduced production of $A\beta_{1-40}$ and $A\beta_{1-42}$ peptides [52]. A_{2A}Rs antagonists and/or caffeine prophylactic and long-term neuroprotective process are suggested to be based on inhibition of reactive oxygen species activity, tau pathology, and Aβ production by neuronal cells [53].

A_{2A}Rs antagonist may also serve as antidepressants, as observed in animal model of antidepressants screening test done by El-Yacoubi et al. [54, 55]. In both tests, A_{2A}Rs antagonists prolong escape-directed behavior. Additionally, potential role as antidepressants was also observed in attenuated behavioral despairs displayed in both tests [55]. The relation between adenosine and depression in preclinical models was obtained from the genetic manipulation model of A_{2A}R. Genetic depletion of A_{2A}Rs resulted in antidepressant-like phenotype in animal models [55]. The

A_{2A}Rs blockade also relieves stress-induced early hippocampal modifications [56]. However, the effect of adenosine neuromodulation system in depression is complex, as it has the ability to modulate several other neurotransmission systems [35].

As addressed in previous paragraphs, A_{2A}R emerges as potential target candidate in various disorders. This is majorly caused by its unique interaction with D2 receptors, a major psychoactive drug target. Important roles of A_{2A}R were also observed in its robust neuroprotective activity, in which it mainly acts in the normalization of glutaminergic synapses, the control of mitochondria-induced apoptosis, and the control of neuroinflammation [35].

4. Current sources of the adenosine A_{2A} antagonist

The treatment of PD currently focuses on symptom management with dopaminergic therapy, such as dopamine precursor L-3,4-dihydroxyphenylalanine (L-DOPA) (in combination with peripheral decarboxylase inhibitors) and dopamine agonists [57]. Although L-DOPA is beneficial in patients with PD, with time, the span of the effect is shortened), the response becomes less probable, and involuntary muscle movements or, in a severe situation, dystonia can emerge [57]. These problems highlight the urgent medical need for an alternative mode of therapeutic intervention that can relieve the symptoms of the disorder while also allowing a decrease in the occurrence of side effects.

Among the non-dopaminergic therapies investigated for the treatment of PD, the adenosine A_{2A} receptor antagonists show very convincingly for two main reasons: their selective and restricted localization in the basal ganglia circuitry and their interaction with dopaminergic receptors. In another word, inhibition of the interaction of adenosine with the A_{2A} receptor may provide a potential treatment for PD.

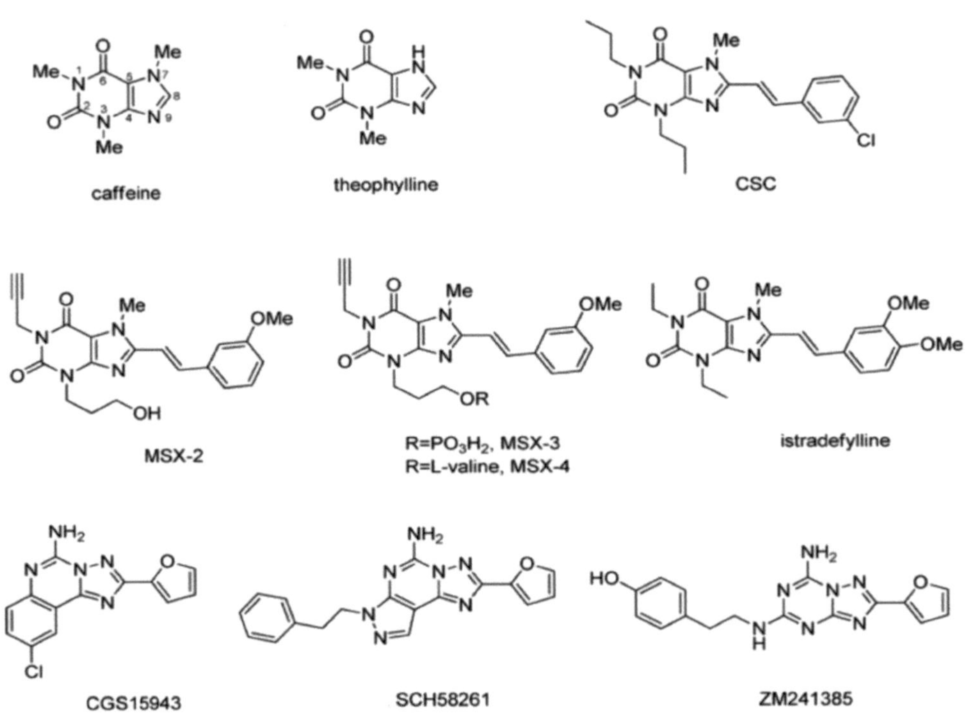

Figure 3.
Adenosine A_{2A} inhibitors.

Many highly selective A_{2A} antagonists, both xanthine and non-xanthine derivatives, have been created, and some of them are being investigated as treatment for subjects with PD in various stage of clinical trials (**Figure 3**) [7, 19, 58–61]. Caffeine as a xanthine derivate is developed as a lead compound for the design of antagonist of adenosine A_{2A} receptor [62]. Experimental model using 1-methyl-4-phenyl-1,2,3,6-tetrahydropyridine (MPTP)-induced parkinsonism is known to be an evidence that caffeine have a protective effect in Parkinson's disease [36, 63]. Some A_{2A} antagonists have progressed to clinical trials by various pharmaceutical companies including istradefylline [59], PBS-509, ST1535 and its metabolite ST4206, tozadenant, V81444, preladenant, and vipadenant [64]. Several studies of novel series of 2-aminoimidazo[4,5-b]pyridine-derivatives [65], arylindenopyrimidine [66], and bicyclic aminoquinazoline derivatives [67] as adenosine A_{2A} antagonists are reported.

Various computational methods were used to study neuroprotective effect from adenosine A_{2A} antagonists such as pharmacophore model [68], QSAR, molecular docking [69–71], and molecular dynamics [72, 73]. Orally bioavailable adenosine A_{2A} receptor antagonists have been studied for its QSAR and pharmacokinetics properties [74].

The study of structure-kinetics relationship (SKR) is done as a complement to a SAR analysis at the adenosine A_{2A} receptor. The series of 24 triazolotriazine derivatives showing a similar binding kinetics to the putative antagonist ZM241385 (4-(2-((7-amino-2-(furan-2-yl)-[1,2,4]triazolo[1,5-a][1,3,5]triazin-5-yl)amino) ethyl)phenol) revealed minor affinity changes, although they varied substantially in their dissociation rates from the receptor [75].

5. Future direction of drug discovery of Parkinson's disease

Various studies have been conducted in the discovery of Parkinson's drugs against the target A_{2A} receptors. The discovery of drugs assisted by computers has accelerated in obtaining lead compounds. Apparently, this method takes a lot of consideration before entering the preclinical and clinical phases. It is because this computational method is more able to describe the answer in preparing the next design. This method can also make various predictions of activities that are difficult to do in the absence of chemical compounds before they are synthesized. In silico prediction of various pharmacokinetic parameters and toxicity can also be done faster. All of these things can provide a better picture of getting a cure for Parkinson's disease.

6. Conclusions

A_{2A} receptors emerge as potential target candidate in various disorders, caused by its unique interaction with D2 receptors, a major psychoactive drug target. Various studies have been conducted in the discovery of Parkinson's drugs against the target A_{2A} receptors. In silico study brings a new approach of study with A_{2A} receptors.

Acknowledgements

This work was supported by *Hibah Publikasi Internasional Terindeks Untuk Tugas Akhir Mahasiswa UI* (PITTA) 2018 by Universitas Indonesia.

Conflict of interest

The authors declare that they have no conflict of interest or involvement with any organization of affiliation.

Author details

Widya Dwi Aryati, Nabilah Nurtika Salamah, Rezi Riadhi Syahdi and Arry Yanuar*
Faculty of Pharmacy, Universitas Indonesia, Depok, Indonesia

*Address all correspondence to: arry.yanuar@ui.ac.id

References

[1] Latini S, Pedata F. Adenosine in the central nervous system: Release mechanisms and extracellular concentrations. Journal of Neurochemistry. 2008;**79**:463-484

[2] Stiles GL. Adenosine receptors. Journal of Biological Chemistry. 1992;**267**:6451-6454

[3] Townsend-Nicholson A, Baker E, Schofield PR, Sutherland GR. Localization of the adenosine A_1 receptor subtype gene (ADORA1) to chromosome 1q32.1. Genomics. 1995;**26**:423-425

[4] Livingston M, Heaney LG, Ennis M. Adenosine, inflammation and asthma? A review. Inflammation Research. 2004;**53**:171-178

[5] Boison D, Singer P, Shen H-Y, Feldon J, Yee BK. Adenosine hypothesis of schizophrenia—Opportunities for pharmacotherapy. Neuropharmacology. 2012;**62**:1527-1543

[6] Fredholm BB, Cunha RA, Svenningsson P. Pharmacology of adenosine A_{2A} receptors and therapeutic applications. Current Topics in Medicinal Chemistry. 2003;**3**:413-426

[7] de Lera Ruiz M, Lim Y-H, Zheng J. Adenosine A_{2A} receptor as a drug discovery target. Journal of Medicinal Chemistry. 2014;**57**:3623-3650

[8] Kull B, Svenningsson P, Fredholm BB. Adenosine A_{2A} receptors are colocalized with and activate golf in rat striatum. Molecular Pharmacology. 2000;**58**:771-777

[9] Torvinen M et al. Adenosine A_{2A} receptor and dopamine D_3 receptor interactions: evidence of functional A_{2A}/D_3 heteromeric complexes. Molecular Pharmacology. 2005;**67**:400-407

[10] Schapira AHV. Present and future drug treatment for Parkinson's disease. Journal of Neurology, Neurosurgery, and Psychiatry. 2005;**76**:1472-1478

[11] Pollack AE, Fink JS. Adenosine antagonists potentiate D2 dopamine-dependent activation of Fos in the striatopallidal pathway. Neuroscience. 1995;**68**:721-728

[12] Lozano AM, Lang AE, Hutchison WD, Dostrovsky JO. New developments in understanding the etiology of Parkinson's disease and in its treatment. Current Opinion in Neurobiology. 1998;**8**:783-790

[13] Savitt JM. Diagnosis and treatment of Parkinson disease: Molecules to medicine. Journal of Clinical Investigation. 2006;**116**:1744-1754

[14] Dauer W, Przedborski S. Parkinson's disease. Neuron. 2003;**39**:889-909

[15] Antonini A, Tolosa E, Mizuno Y, Yamamoto M, Poewe WH. A reassessment of risks and benefits of dopamine agonists in Parkinson's disease. Lancet Neurology. 2009;**8**:929-937

[16] Yamamoto M, Schapira AH. Dopamine agonists in Parkinson's disease. Expert Review of Neurotherapeutics. 2008;**8**:671-677

[17] Olanow CW, Stocchi F. COMT inhibitors in Parkinson's disease: Can they prevent and/or reverse levodopa-induced motor complications? Neurology. 2004;**62**:S72-S81

[18] Gordin A, Brooks DJ. Clinical pharmacology and therapeutic use of COMT inhibition in Parkinson's disease. Journal of Neurology. 2007;**254**:IV37-IV48

[19] Shook BC, Jackson PF. Adenosine A_{2A} receptor antagonists and Parkinson's

disease. ACS Chemical Neuroscience. 2011;**2**:555-567

[20] Schwarzschild MA, Agnati L, Fuxe K, Chen J-F, Morelli M. Targeting adenosine A_{2A} receptors in Parkinson's disease. Trends in Neurosciences. 2006;**29**:647-654

[21] Salamone JD. Facing dyskinesia in Parkinson disease: Nondopaminergic approaches. Drugs Future. 2010;**35**:567

[22] Rosin DL, Robeva A, Woodard RL, Guyenet PG, Linden J. Immunohistochemical localization of adenosine A_{2A} receptors in the rat central nervous system. Journal of Comparative Neurology. 1998;**401**:163-186

[23] Fredholm BB, Svenningsson P. Striatal adenosine A_{2A} receptors—Where are they? What do they do? Trends in Pharmacological Sciences. 1998;**19**:46-47

[24] Ishiwata K et al. First visualization of adenosine A_{2A} receptors in the human brain by positron emission tomography with [^{11}C]TMSX. Synapse. 2005;**55**:133-136

[25] Svenningsson P, Hall H, Sedvall G, Fredholm BB. Distribution of adenosine receptors in the postmortem human brain: An extended autoradiographic study. Synapse. 1997;**27**:322-335

[26] Fink JS et al. Molecular cloning of the rat A_2 adenosine receptor: Selective co-expression with D_2 dopamine receptors in rat striatum. Molecular Brain Research. 1992;**14**:186-195

[27] Chen J-F et al. Neuroprotection by caffeine and A_{2A} adenosine receptor inactivation in a model of Parkinson's disease. Journal of Neuroscience. 2001;**21**:RC143-RC143

[28] Grondin R et al. Antiparkinsonian effect of a new selective adenosine A_{2A} receptor antagonist in MPTP-treated monkeys. Neurology. 1999;**52**:1673-1673

[29] Ongini E et al. Dual actions of A_{2A} adenosine receptor antagonists on motor dysfunction and neurodegenerative processes. Drug Development Research. 2001;**52**:379-386

[30] Ikeda K, Kurokawa M, Aoyama S, Kuwana Y. Neuroprotection by adenosine A_{2A} receptor blockade in experimental models of Parkinson's disease. Journal of Neurochemistry. 2002;**80**:262-270

[31] Kanda T et al. Combined use of the adenosine A_{2A} antagonist KW-6002 with l-DOPA or with selective D1 or D2 dopamine agonists increases antiparkinsonian activity but not dyskinesia in MPTP-treated monkeys. Experimental Neurology. 2000;**162**:321-327

[32] Hauser RA, Hubble JP, Truong DD. Randomized trial of the adenosine A_{2A} receptor antagonist istradefylline in advanced PD. Neurology. 2003;**61**:297-303

[33] Fuxe K et al. Adenosine-dopamine interactions in the pathophysiology and treatment of CNS disorders. CNS Neuroscience and Therapeutics. 2010;**16**:e18-e42

[34] Cunha RA. Neuroprotection by adenosine in the brain: From A_1 receptor activation to A_{2A} receptor blockade. Purinergic Signal. 2005;**1**:111-134

[35] Gomes CV, Kaster MP, Tomé AR, Agostinho PM, Cunha RA. Adenosine receptors and brain diseases: Neuroprotection and neurodegeneration. Biochimica et Biophysica Acta (BBA)—Biomembranes. 2011;**1808**:1380-1399

[36] Kalda A, Yu L, Oztas E, Chen J-F. Novel neuroprotection by caffeine and adenosine A_{2A} receptor antagonists

in animal models of Parkinson's disease. Journal of the Neurological Sciences. 2006;**248**:9-15

[37] Yamada K, Kobayashi M, Kanda T. Chapter Fifteen - Involvement of Adenosine A_{2A} Receptors in Depression and Anxiety. International Review of Neurobiology. 2014;**119**:373-393. DOI: 10.1016/B978-0-12-801022-8.00015-5

[38] Sako W, Murakami N, Motohama K, Izumi Y, Kaji R. The effect of istradefylline for Parkinson's disease: A meta-analysis. Scientific Reports. 2017;**7**:18018

[39] Isaacson S et al. Efficacy and safety of istradefylline in moderate to severe Parkinson's disease: A phase 3, multinational, randomized, double-blind, placebo-controlled trial (i-step study). Journal of the Neurological Sciences. 2017;**381**:351-352

[40] Franco R, Navarro G. Adenosine A_{2A} receptor antagonists in neurodegenerative diseases: Huge potential and huge challenges. Frontiers in Psychiatry. 2018;**9**(1-5)

[41] Chen J-F, Pedata F. Modulation of ischemic brain injury and neuroinflammation by adenosine A_{2A} receptors. Current Pharmaceutical Design. 2008;**14**:1490-1499

[42] Gao Y, Phillis JW. CGS 15943, An adenosine A2 receptor antagonist, reduces cerebral ischemic injury in the Mongolian gerbil. Life Sciences. 1994;**55**:PL61-PL65

[43] Pedata F et al. Adenosine A_{2A} receptors modulate acute injury and neuroinflammation in brain ischemia. Mediators of Inflammation. 2014;**805198**:2014

[44] Kalaria RN et al. Alzheimer's disease and vascular dementia in developing countries: Prevalence, management, and risk factors. The Lancet Neurology. 2008;**7**:812-826

[45] Lesne S. Toxic oligomer species of amyloid-β in Alzheimer's disease, a timing issue. Swiss Medical Weekly. 2014;**298**:789-791

[46] Cunha RA, Agostinho PM. Chronic caffeine consumption prevents memory disturbance in different animal models of memory decline. Journal of Alzheimer's Disease. 2010;**20**:S95-S116

[47] Ribeiro JA, Sebastião AM, de Mendonça A. Adenosine receptors in the nervous system: Pathophysiological implications. Progress in Neurobiology. 2002;**68**:377-392

[48] Maia L, de Mendonça A. Does caffeine intake protect from Alzheimer's disease? Europen Journal of Neurology. 2002;**9**:377-382

[49] Ritchie K et al. The neuroprotective effects of caffeine: A prospective population study (the Three City Study). Neurology. 2007;**69**:536-545

[50] Santos C, Costa J, Santos J, Vaz-Carneiro A, Lunet N. Caffeine intake and dementia: systematic review and meta-analysis. Journal of Alzheimer's Disease. 2010;**20**:S187-S204

[51] Smith AP. Caffeine, cognitive failures and health in a non-working community sample. Human Psychopharmacology Clinical and Experimental. 2009;**24**:29-34

[52] Arendash GW et al. Caffeine protects Alzheimer's mice against cognitive impairment and reduces brain β-amyloid production. Neuroscience. 2006;**142**:941-952

[53] Marzagalli R, Castorina A. The seeming paradox of adenosine receptors as targets for the treatment of Alzheimer's disease: Agonists or antagonists? Neural Regeneration Research. 2015;**10**:205

[54] Cantwell R, Cox JL. Psychiatric disorders in pregnancy and the

puerperium. Current Obstetrics & Gynaecology. 2006;**16**:14-20

[55] El Yacoubi M et al. Adenosine A_{2A} receptor antagonists are potential antidepressants: evidence based on pharmacology and A_{2A} receptor knockout mice. British Journal of Pharmacology. 2001;**134**:68-77

[56] Cunha GMA, Canas PM, Oliveira CR, Cunha RA. Increased density and synapto-protective effect of adenosine A_{2A} receptors upon sub-chronic restraint stress. Neuroscience. 2006;**141**:1775-1781

[57] Olanow CW et al. Levodopa in the treatment of Parkinson's disease: Current controversies. Movement Disorders. 2004. DOI: 10.1002/mds.20243

[58] Armentero MT et al. Past, present and future of A_{2A} adenosine receptor antagonists in the therapy of Parkinson's disease. Pharmacology & Therapeutics. 2011. DOI: 10.1016/j.pharmthera.2011.07.004

[59] Jenner P. Istradefylline, a novel adenosine A_{2A} receptor antagonist, for the treatment of Parkinson's disease. Expert Opinion on Investigational Drugs. 2005. DOI: 10.1517/13543784.14.6.729

[60] Pinna A. Novel investigational adenosine A_{2A} receptor antagonists for Parkinson's disease. Expert Opinion on Investigational Drugs. 2009. DOI: 10.1517/13543780903241615

[61] Hickey P, Stacy M. Adenosine A_{2A} antagonists in Parkinson's disease: What's next? Current Neurology and Neuroscience Reports. 2012;**12**:376-385

[62] Petzer JP, Petzer A. Caffeine as a lead compound for the design of therapeutic agents for the treatment of Parkinson's disease. Current Medicinal Chemistry. 2015. DOI: 10.2174/0929867322666141215160015

[63] Munoz DG, Fujioka S. Caffeine and Parkinson disease: A possible diagnostic and pathogenic breakthrough. Neurology. 2018. DOI: 10.1212/WNL.0000000000004898

[64] Pinna A. Adenosine A_{2A} receptor antagonists in Parkinson's disease: Progress in clinical trials from the newly approved istradefylline to drugs in early development and those already discontinued. CNS Drugs. 2014;**28**:455-474

[65] McGuinness BF et al. Discovery of 2-aminoimidazopyridine adenosine A_{2A} receptor antagonists. Bioorganic & Medicinal Chemistry Letters. 2010;**20**:6845-6849

[66] Atack JR et al. JNJ-40255293, a novel adenosine A_{2A}/A_1 antagonist with efficacy in preclinical models of Parkinson's disease. ACS Chemical Neuroscience. 2014;**5**:1005-1019

[67] Zhou G et al. Bioorganic & medicinal chemistry letters discovery of aminoquinazoline derivatives as human A_{2A} adenosine receptor antagonists. Bioorganic & Medicinal Chemistry Letters. 2016;**26**:1348-1354

[68] Khanfar MA, Al-Qtaishat S, Habash M, Taha MO. Discovery of potent adenosine A_{2A} antagonists as potential anti-Parkinson disease agents. Non-linear QSAR analyses integrated with pharmacophore modeling. Chemico-Biological Interactions. 2016;**254**:93-101

[69] Anighoro A, Bajorath J. Binding mode similarity measures for ranking of docking poses: A case study on the adenosine A_{2A} receptor. Journal of Computer-Aided Molecular Design. 2016. DOI: 10.1007/s10822-016-9918-z

[70] Yang X et al. A covalent antagonist for the human adenosine A_{2A} receptor. Purinergic Signal. 2017. DOI: 10.1007/s11302-016-9549-9

[71] Jaiteh M et al. Docking screens for dual inhibitors of disparate drug targets for Parkinson's disease. Journal of Medicinal Chemistry. 2018. DOI: 10.1021/acs.jmedchem.8b00204

[72] Sabbadin D et al. Bridging molecular docking to membrane molecular dynamics to investigate GPCR-ligand recognition: The human A_{2A} adenosine receptor as a key study. Journal of Chemical Information and Modeling. 2013. DOI: 10.1021/ci400532b

[73] Caliman AD, Swift SE, Wang Y, Miao Y, McCammon JA. Investigation of the conformational dynamics of the apo A_{2A} adenosine receptor. Protein Science. 2015. DOI: 10.1002/pro.2681

[74] Basu S et al. Design, synthesis of novel, potent, selective, orally bioavailable adenosine A_{2A} receptor antagonists and their biological evaluation. Journal of Medicinal Chemistry. 2017. DOI: 10.1021/acs.jmedchem.6b01584

[75] Guo D et al. Binding kinetics of ZM241385 derivatives at the human adenosine A_{2A} receptor. ChemMedChem. 2014. DOI: 10.1002/cmdc.201300474

Chapter 4

Adrenergic Receptors as Pharmacological Targets for Neuroinflammation and Neurodegeneration in Parkinson's Disease

Monika Sharma and Patrick M. Flood

Abstract

Inflammation is a key component of the dopaminergic neurodegeneration seen in progressive Parkinson's disease (PD). The presence of activated glial cells, the participation of innate immune system, increased inflammatory molecules such as cytokines and chemokines, and increased oxidative stress and reactive oxygen species are the main neuroinflammatory characteristics present in progressive PD. Therapeutic targets which suppress pro-inflammatory responses by glial cells (mainly microglia) have been shown to be effective treatments for slowing or eliminating the progressive degeneration of neurons within the substantia nigra. In this chapter, we will detail a specific anti-inflammatory therapy using agonists to β2-adrenergic receptors that have been shown to be effective treatments for models of dopaminergic neurodegeneration and that have had efficacy in patients with progressive PD. We will also detail the possible molecular mechanisms of action of this therapeutic in stopping or reversing inflammation within the CNS.

Keywords: β2-adrenergic receptor, Parkinson's disease, microglia, neuroinflammation

1. Introduction

There are a number of neurological disorders that fall under the umbrella of neurodegeneration, with the major ones including Alzheimer's disease (AD), Parkinson's disease (PD), Huntington's disease (HD), amyotrophic lateral sclerosis (ALS), frontotemporal dementia (FTD), spinal cord injury (SCI), and others. Currently, there are no generally effective treatments available to slow down or reverse the debilitating effects of these diseases, and the long-term effects of these diseases are the progressive degeneration and death of neurons. A majority of the neurodegenerative diseases are linked with inflammation in CNS [1], and the presence of activated glial cells, infiltration and activation of adaptive and innate immune cells, increased presence of inflammatory molecules such as cytokines and chemokines, and increased oxidative stress and reactive oxygen species (ROS) are the main neuroinflammatory characteristics present in lesions associated with these neurodegenerative disorders. Recent approaches found to be effective in the treatment of Parkinson's disease involve the use of anti-inflammatory agents and cytokines such as agonists to the β2-adrenergic receptors (β2-AR) to inhibit neuroinflammation and the progression of dopaminergic neurodegeneration. In this chapter, we will address the current understanding of therapeutic approaches targeting neuroinflammation linked with PD and the use of β2-AR agonists as an effective treatment for PD.

2. Parkinson's disease: a chronic neurodegenerative and neuroinflammatory disease

Parkinson's disease (PD) is a progressive neurodegenerative disorder which leads to impaired motor skills. The major pathological feature of PD is the degeneration of dopaminergic (DA) neurons which project from substantia nigra (SN) to the striatum in the midbrain (nigro-striatal pathway) [2]. Another neuropathological feature of PD is the cytoplasmic inclusion of misfolded α-synuclein protein in degenerating dopaminergic neurons called Lewy bodies [3]. The primary motor symptoms of PD, such as tremor, rigidity, and bradykinesia, are caused by inadequate formation and neurotransmission of dopamine within the nigro-striatal pathway [4, 5]. Dementia is reported in 28% of PD cases with the prevalence rising to 65% in those aged 85 years and above. Patients with PD also show non-motor-related symptoms such as olfactory deficits, depression, cognitive deficits, sleep disorders, and autonomic dysfunction [6]. The majority of PD cases are idiopathic Parkinson's, and the disease mechanism that ultimately causes idiopathic PD is largely unknown. In the remainder of the cases of PD, about 10–15% of patients do have a family history and those patients are referred to as having the *familial* form of PD. For these patients, their PD appears to be caused by a mutation in one of a few selected genes (such as *SNCA*, *Parkin*, *LRRK2*, *DJ-1*, etc.) [7, 8]. Although the etiology of the idiopathic form of the disease remains elusive, there are some risk factors associated with the development of the disease. These risk factors include exposure to environmental toxins, severe cranial trauma, systemic or localized infections, and inherited genetic risk factors. These genetic and nongenetic risk factors have the potential to initiate neurodegeneration and subsequent chronic inflammation in the brain which eventually contributes to the pathophysiology of PD [9]. In addition, several cellular and molecular pathways such as oxidative stress [10], proteosomal dysfunction [11], excitotoxicity [12], and mitochondrial dysfunction [13] have also been identified which contributes to neuronal death.

The presence of activated glial cells, increased inflammatory molecules such as cytokines/chemokines, and increased oxidative stress and ROS are the main neuroinflammatory characteristics present in PD [14]. PD is now not only characterized as loss of DA-neurons and motor impairment, but also recognized to have an inflammatory component which plays a crucial role in the progression of the disease. Several inflammatory mediators such as TNF-α, IL-1β, ROS, and nitric oxide (NO), released from nonneuronal cells exacerbate the disease pathology [3, 15]. It has been suggested that α-synuclein released from dying neurons also activate the microglia via TLR2 activation [16]. Furthermore, the elevated levels of inflammatory cytokines such as TNF-α, IL-1β, and IL-6 have been reported in serum, cerebrospinal fluid (CSF), and striatum of PD patients [17]. The influx of peripheral macrophages has been reported in brains of patients with PD, but the role of these cells in disease pathology remains to be tested [18]. Additionally, activation and increased number of glial cells and infiltrating peripheral lymphocytes such as cytotoxic CD4+ and CD8+ cells in SN also support the role of adaptive immunity in the etiology of the disease [8]. Overall, these studies and others suggest the contribution of the immune system in the pathophysiology of PD.

3. Microglial activation and neuroinflammation in PD

Microglia originate from erythromyeloid progenitors in the yolk sac which migrate and differentiate during development to form the central nervous system (CNS). Fully differentiated microglial cells are also considered to be the resident macrophages of the CNS [19], although some phenotypic and functional differences between microglia and macrophages have been found [20]. Growing evidence suggests that the activation of microglia in CNS plays an important role in the pathogenesis of PD. It is not well understood how microglia activation is either beneficial or detrimental to the neuron or how microglial activity is regulated. It has been found that microglial activation is required for neuronal survival by the removal of toxic substances through innate immunity [21]. On the other hand, it has been found that *over*-activated microglial cells are detrimental and neurotoxic [22]. Research studies of post-mortem brain tissue from patients with PD and related parkinsonian syndromes suggest the presence of activated microglia around degenerating DA-neurons in the SN [23] and these activated microglia are not only limited to the SN but also present in extended brain areas such as hippocampus, putamen, trans-entorhinal cortex, cingulate cortex, and temporal cortex [24]. Imaging of activated microglia in the striatum could be used as a biomarker for detecting neuroinflammation in neurodegenerative parkinsonian disorders [25]. The resting microglia switches to an activated microglia phenotype in response to pathogen invasion or release of toxic or inflammatory mediators and thereby promotes an inflammatory response [1]. Once activated, microglial cells produce a wide range of inflammatory mediators which serve to initiate an innate immune response or glial cell-propagated inflammation termed as neuroinflammation [26]. Also, the degenerating DA-neurons release many toxic factors that activate microglia and these degenerating neurons are vulnerable to inflammatory insult. Degenerating neurons will co-localize or attract an even larger population of microglia in the SN [27]. Collectively, these activated microglia and damaged neurons form a repetitive and vicious cycle that leads to chronic inflammation and continued extensive DA neurodegeneration over time, leading to the progression of PD [27]. These findings confirm neuroinflammation as a pivotal process in the progression of neurodegenerative disorders and the central role of microglia in this process [22]. Targeting neuroinflammatory pathways within microglia could be a significant step in the development of new therapeutics for neurodegenerative diseases, including PD.

4. Therapies targeting neurodegeneration/neuroinflammation in PD

Treatment for PD normally involves medications such as Levodopa to enhance the dopamine levels and deal with movement symptoms [28]. While none of our current treatments are able to stop the disease, medication and surgery can be helpful for managing the symptoms [29]. These treatments work well in patients initially, but they are also associated with unwanted side-effects and reduced efficacy over time [30]. On the other hand, many studies suggest that inflammatory mediators such as TNF, PGE_2, NO, free radicals, and other immune mediators play role in the pathogenesis of PD and degeneration of dopamine-producing neurons and that targeting these mediators can be an effective treatment for PD. This opens up the potential of using anti-inflammatory drugs as an effective and long-term treatment in PD. These anti-inflammatory drugs can act by arresting the disease onset (primary prevention) or by interrupting or even reversing the disease progression (secondary prevention). Epidemiological and observational studies suggest that the use of anti-inflammatory drugs lower the risk of developing PD [31]. Observations which demonstrated that inflammation in SN plays a role in PD have led many investigators to initially consider the potential use of both steroidal and nonsteroidal anti-inflammatory drugs for the treatment of PD. Steroidal anti-inflammatory drugs (SAIDs), such as dexamethasone, have shown neuroprotective effects in LPS-induced neurotoxicity in the SN in LPS models of PD [32]. Nonsteroidal anti-inflammatory drugs (NSAIDs) have also been used as analgesics and antipyretics to suppress the adverse effects of inflammation [33]. The neuroprotective effects of Ibuprofen have been studied in PD pathogenesis and these studies demonstrate the protective effect on dopaminergic neurons against glutamate toxicity *in vitro* [34, 35]. Previously, we have established several therapies targeting neuroinflammation and neurodegeneration in an animal model of PD and these therapies include D-morphinan-related compounds [36], anti-inflammatory cytokines such as TGF-β (transforming growth factor-beta) [37] and IL-10 [38, 39], IKK (inhibitor of kappa B (IκB) kinase) inhibitors [40], NADPH (nicotinamide adenine dinucleotide phosphate) oxidase inhibitors [41], and β2-AR (beta 2-adrenergic receptor) agonists [42, 43].

We have conducted a number of experiments using different classes of anti-inflammatory compounds to determine their efficacy in preventing dopaminergic neurotoxicity by activated microglial cells both *in vitro* and *in vivo*. First, it was found that morphinan compounds and their stereoisomers (L-morphine and its D stereo enantiomers) can inhibit microglial activation and LPS- or MPP^+-induced neurotoxicity in rat primary mesencephalic cultures. We and others observed that several dextrorotatory isomers of morphine compounds, including D-morphine, dextromethorphan, and sinomenine, showed neuroprotective effects against LPS and MPP+ (1-methyl-4-phenylpyridinium) which were mediated through the inhibition of microglial PHOX activity [36, 44, 45]. Furthermore, these studies also suggest that these morphinan compounds bind to the catalytic subunit of PHOX, inhibit its activity, and reduce the production of superoxide and other pro-inflammatory cytokines [44]. In another set of studies using a different anti-inflammatory approach, a specific inhibitor of IKK-β (IkappaB kinase-beta) protects dopaminergic neurons against LPS-induced neurotoxicity both *in vitro* and *in vivo* through inhibition of NF-κB activation, resulting in the decreased production of ROS and inflammatory cytokines [40]. We have also developed therapies targeting neuroinflammation in PD models by using anti-inflammatory cytokines such as IL-10 and TGF-β1, and found that treatment with IL-10 on rat mesencephalic neuron-glia culture protects against LPS-induced neurotoxicity via suppression of pro-inflammatory mediators and superoxide production [38]. Similarly, the neuroprotective effect of TGFβ1 is primarily due to its ability to inhibit ERK phosphorylation, the serine phosphorylation on $p47^{phox}$, and the production of ROS from microglia during activation by LPS [37].

5. Adrenergic receptors

One of the most potent and successful therapeutic treatments for inflammation-mediated dopaminergic neurotoxicity is the use of long-acting agonists to the β2-AR. Adrenergic receptors (AR) are seven-transmembrane proteins that serve as adrenoreceptors for catecholamines such as norepinephrine and epinephrine on multiple cell types, and cells within the CNS that express AR include neurons, immune cells, and

Receptor Type	Tissue Distribution	Mechanism of Action	Agonist Potency	Physiological Effects	Agonist	Antagonist
α1	Vascular Smooth Muscles, Visceral smooth Muscles	Gq-protein coupled activates Phospholipase C, IP3+DAG	Epi ≥ NE >> Iso	Smooth muscle contractions, Gluconeogenesis, Vasoconstriction	Norepinephrine, Phenylephrine, Methoxamine	Doxazosin, Phentolamine, Prazosin
α2	Pre-synaptic terminals, pancreas, platelets, Ciliary epithelium, Salivary Glands	Gi-protein coupled inhibits Adenyl cyclase	Epi ≥ NE >>Iso	Inhibits release of Neurotransmitter	Clonidine, Monoxidine	Yohimbine, Idazoxan, Tolazoline
β1	Heart, Kidney, some pre-synaptic terminals	Gs-protein coupled activates Adenyl cyclase +PKA	Iso > Epi ≥ NE	Increase heart rate and Renin secretion	Isoproterenol, Norepinephrine, Dobutamine	Propranolol, Metoprolol, Atenolol
β2	Visceral smooth muscles, Bronchioles, Liver, Skeletal Muscles	Gs-protein coupled activates Adenyl cyclase +PKA, Ca-channels	Iso > Epi >> NE	Vasodilation, Bronchodilation, Inhibits insulin secretion	Isoproterenol, Salbutamol, Salmeterol, Albuterol, Formoterol, Terbutaline, Levalbuterol	Propranolol, ICI-118,551, Nadolol, Butoxamine
β3	Adipose Tissue	Gs-protein coupled activates Adenyl cyclase +PKA	Iso = NE > Epi	Increase lipolysis	Isoproterenol, Amibegron, Solabegron	SR59230A

NE: Norepinephrine, **Epi:** Epinephrine and **Iso:** Isoproterenol

Table 1.
Characteristics of adrenergic receptors.

astrocytes. Pharmacological classification of the adrenergic receptor was first introduced in 1948 and broadly classified as α and β adrenergic receptors [46] by Ahlquist. The classification was based on the order of potency and specificity of natural and synthetic agonist and blocking agents. The α-AR response corresponds to mainly excitatory response, while β-AR responses were correlated mainly with the inhibitory response. The α-AR response showed the order of potency: norepinephrine > epinephrine > isoproterenol and β-AR-mediated response exhibited order of potency: isoproterenol > epinephrine > norepinephrine [47, 48]. After the discovery of new drugs which have a high affinity to adrenergic receptors, these receptors were sub-classified. α-AR were subdivided into α1 and α2 adrenergic receptors [49]. Further studies subdivided β-AR into β1 and β2 which are normally present on immune cells, cardiac muscles, and airway smooth muscles, respectively [50]. A third β-AR, now called as β3-AR was identified on adipose tissues [51]. Tissue distribution, physiological effects, mechanism of action, and the major agonists/antagonists of ARs are summarized in **Table 1**. Pharmacological compounds that serve as short, long, and ultra-long-acting agonists for these receptors have now been developed, and they are normally thought to stimulate adrenergic receptors by four different mechanisms: (1) by direct receptor binding, the most common mechanism where drugs activate peripheral adrenergic receptors via direct binding to receptor and mimic the actions of endogenous agonists (NE, epinephrine), (2) by ameliorating NE release, where drugs act on sympathetic nerve terminals and results into NE release, (3) by inhibition of NE reuptake, where these drugs can cause NE to accumulate within synaptic gaps at sympathetic nerve terminals, (4) by blockade of NE inactivation where drugs inhibit the activity of monoamine oxidase (MAO) which inhibits the activity of monoamines such as NE and dopamine [52].

6. General properties of β2-adrenergic receptors: a G-protein-coupled receptor

6.1 Structure

The β2-ARs belong to a diverse superfamily of human cell surface seven trans-membrane receptors for hormones and neurotransmitters called G-protein-coupled receptors (GPCRs). GPCRs are divided into six classes on basis of sequence homology: class A (Rhodopsin-like), class B (Secretin receptor family), class C (Metabotropic glutamate), class D (Fungal mating pheromone receptor), class E (Cyclic AMP receptor), and class F (Frizzled/smoothened) [53]. GPCRs are one of the most extensively studied proteins for the development of pharmaceutical drugs and target for approximately 50% of the marketed pharmaceutical drugs [54]. The adrenergic receptor family belongs to the rhodopsin-like subfamily, the largest class of the GPCR. The β2-AR is an intron-less gene is present on the long arm of chromosome 5 (5q31) and encodes for 413 amino acid polypeptide of 46kD [55]. Similar to all GPCRs, β2-AR is composed of seven transmembrane spanning α-helices with an intracellular C-terminus and an extracellular N-terminus. The β2-AR was the first GPCR to be cloned [56] and the first GPCR structure to be solved [57]. The β2-AR has been studied extensively and also serves as a model system for investigating the regulation and signal transduction of GPCRs. The study of the 3D protein structure of this family of GPCRs took a giant leap forward when rhodopsin was first crystallized in 2000 and this crystalline structure has been used as an important template for modeling other GPCRs in this family [58]. The crystalline structure of human β2-AR was not solved until 2007, when a nonactive structure of β2-AR was identified [57]. Post-translational modifications such as glycosylation, pamitoylation, disulfide bond formation, and phosphorylation have now been found to affect receptor functions. Interestingly, β2-AR is glycosylated at amino acid 6, 15, and 187 which is important for the trafficking of the β2-AR from the endoplasmic reticulum to the plasma membrane [59]. Mutation in these sites also results in reduced expression of receptor on the cell membrane, suggesting a role for glycosylation in cell surface expression [60]. Conversely, the cysteine amino acid in the cytoplasmic tail at position 341 is palmitoylated, and is now found to be an important residue for the adequate coupling of the receptor to the G_s-protein [61]. Finally, β2-ARs have disulfide bonds which are essential for agonist binding and also for maintaining their tertiary structure [62].

6.2 Localization

Adrenergic receptors are widely distributed on human body organs and regulate physiologic functions such as bronchodilation [63], vasodilation, glycogenolysis in the liver, and relaxation of uterine and bladder muscles [64]. The human β2-AR are widely expressed not only on airway smooth muscles, but also on the wide variety of cells such as epithelial cells, endothelial cells, brain cells, and immune cells including mast cells, macrophages, adaptive immune cells, and eosinophils [65]. The expressions of β1- and β2-AR have also been found on microglial cells, suggesting that microglia, the brain's resident immune cell, is predominantly regulated by NE since NE is the predominant catecholamine in the CNS. Conversely, peripheral immune cells such as macrophages and T cells, which also express high levels of β1 and β-2 AR, are thought to be regulated primarily by epinephrine [66].

6.3 β2-AR activation and signaling pathways in inflammation

Activation of adrenergic receptors could result into both pro- and anti-inflammatory actions, depending on certain parameters such as the type of cell, duration of ligand exposure to the receptor, and type of the adrenergic receptor [67]. It is the diversity of the β2-AR that leads to the complexity of signaling mechanisms and to this duality of function. Activation of β2-AR by receptor agonists initiate intracellular signaling pathways that function either via G-proteins or through β-arrestins. Like other GPCR, β2-AR can activate either canonical (traditional) or noncanonical (nontraditional) signal transduction pathway. In the canonical pathway, similar to a typical GPCR the β2-AR signals via a heterotrimeric G-protein complex, and when the receptor is coupled to inactive GDP-bound G-protein, it appears to have high affinity to the agonist or ligand. After ligand binding, the transmembrane domains of the receptor undergo conformational change with the exchange of GDP to GTP. Further, this conformational change reduces the affinity of the ligand to its receptor, increasing the possibility of retraction of ligand from the receptor, thereby preventing the over-activation of G-protein. This provides evidence that β2-AR appear to oscillate between an active and inactive form under normal conditions. After the exchange of GDP to GTP, the G_{α}-subunit dissociates from $G_{\beta\gamma}$-subunit which remains associated with plasma membrane and the G_{α}-subunit activates effector proteins. The downstream signaling of this process normally results in the production of intracellular second messengers which further activates the cAMP-PKA-mediated intracellular signaling pathway. The activated β2-AR binds with the α-subunit of the G-protein together with a guanosine triphosphate (GTP) molecule. Further, the receptor coupled with adenylate cyclase (AC) which catalyzes the conversion of ATP into cAMP (a second messenger for β2-AR) by hydrolysis of GTP into GDP. The cAMP activates and regulates protein kinase A (PKA) which further mediates the transcription of genes and degradation of cAMP by phosphodiesterase (PDE) leading to termination of signaling [68].

Earlier it was determined that β2-AR exhibits their inhibitory signals in immune cells via the canonical (PKA) signaling pathway. It has now been found that GPCR can also signal through a noncanonical pathway in addition to their classical signaling pathway [69]. Activation through the noncanonical signaling pathway is cell type dependent and G-protein independent, but rather the G-protein-coupled receptor kinases (GRKs) and β-arrestins are involved in activation of this noncanonical signaling pathway. Various types of GRKs phosphorylate specifically serine and threonine at C-terminal of the β2-AR which further determines whether receptors undergo desensitization or initiate noncanonical signaling [70]. For example, phosphorylation of receptor by GRK5/6 initiates β-arrestin-mediated noncanonical signaling, while phosphorylation by GRK2 leads to β-arrestin-mediated desensitization of the receptor [71]. During noncanonical signaling, β-arrestin2 couples β2-AR to MAPK signaling pathways which induces activation of transcription factors and allows their nuclear translocation. Activation of β2-AR with high agonist concentration can lead to sustained activation of ERK1/2 via β-arrestin2. This explains why β2-AR activation can either enhance or suppress the proliferation of immune cells and cytokine production particularly at a high concentration of agonists [67, 72]. Studies suggest that during inflammatory conditions immune cells can switch from canonical to the noncanonical pathway [67, 68]. Engagement of β2-AR receptors by agonists can result in immunomodulatory actions. Depending on the type of immune stimuli and timing of β2-AR activation relative to immune activation, β2-AR stimulation can positively or negatively regulate the response of immune activator [67, 73]. The initial data obtained in animal models of dopaminergic neurotoxicity suggests that the primary immunomodulatory mechanism of β2-AR activation that regulates CNS inflammation in microglial cells occurs through the noncanonical β-arrestin2 pathway of activation.

7. β2-agonists

β-agonists are a group of pharmaceutical compounds or sympathomimetic drugs that mimic the effects of endogenous catecholamines such as epinephrine, norepinephrine, and dopamine. These drugs do not comprise a similar structure to catecholamines but still directly or indirectly activate the β2-adrenergic receptor. The first β-agonist was used around 5000 years ago in Chinese medicine where an ephedrine containing plant, Ma-huang, was used to treat respiratory problems [74]. Further research in the twentieth century has led to increased use of β-agonists for the treatment of respiratory diseases. The first β2-AR selective agonist, Salbutamol was synthesized by Glaxo in 1968 [75]. Later, the same team at Glaxo modified Salbutamol into Salmeterol with long-lasting effects and reduced side effects. Recently, they have synthesized β2-agonists with ultra-long-lasting effects such as Indacaterol [76]. After successful trials, these β2-agonists were approved by the US Food and Drug Administration (FDA) for the treatment of respiratory diseases such as asthma and chronic obstructive pulmonary disease (COPD). Since 1968, a number of companies have labored to develop β2-AR agonists, and some have now been commercialized for use in the treatment of COPD. A list of some of these agonists is given below and in **Table 1**.

7.1 Classification of β2-agonists

A pharmacogenetic study of β2-agonists has summarized the relationship between polymorphisms in the β2-adrenoreceptor (*ADRB2*) gene and the effects of select β2-agonists [77]. Two hypotheses aim to account for the differences in functioning and *in vivo* half-lives of these compounds: exosite/exoreceptor or plasmalemma diffusion microkinetics. Briefly, the exosite hypothesis focuses on the ability of the side-chain of these compounds to interact with a distinct site on the receptor such that it allows the active component to "swing back-and-forth" to activate the receptor. The plasmalemma diffusion microkinetic hypothesis suggests that high concentrations of agonists are achieved in close proximity to the receptor and allows for a longer duration of action [78]. Both of these hypotheses require further investigation and need to be studied within the CNS. Depending upon their mechanism and duration of action, all β2-agonists are grouped into three major classes: short-acting, long-acting, and ultra-long-acting β2-agonists.

7.1.1 Short-acting β-agonists (SABA)

These drugs are mostly hydrophilic in nature, access the active site of β-AR directly from the aqueous extracellular area and show the fast onset of action [79]. These SABAs bind to the receptor for short time; therefore, their duration of action is short. Some of the more common SABAs include Salbutamol (Ventolin), Albuterol (AccuNeb), Pirbuterol (Maxair), and Levalbuterol (Xopenex).

7.1.2 Long-acting β-agonists (LABA)

These drugs are a frontline treatment for COPD, and usually prescribed alone or in combination with inhaled corticosteroids. LABAs are lipophilic in nature and taken up by cell membrane as a reservoir, progressively seep out and interact with the active site of the receptor [79]. They diffuse in the plasma membrane, where they interact with the active site of the β2-AR which allows for the close proximity with the receptor and longer duration of action. The onset of action of these drugs is slower as compared to SABAs, but the duration of action is prolonged thereby, called as LABAs. The duration of action is also dependent on the concentration of the agonist. Salmeterol, Salmeterol with an inhaled corticosteroid, Formoterol, and Formoterol with an inhaled corticosteroid are commercially available LABAs and used in medication for asthma and COPD [80].

7.1.3 Ultra-long-acting β-agonists

These agonists are also lipophilic in nature and onset of action is similar to LABAs, but the duration of action lasts longer than LABAs. Vilanterol with an inhaled corticosteroid and Indacaterol are ultra-LABAs, approved by FDA for the treatment of COPD [81].

8. β2-adrenergic receptors agonists in neuroprotection

The majority of adrenergic neurons are present in brainstem locus coeruleus (LC) nuclei, which is a predominant site for the production of norepinephrine (NE) in the brain. LC neurons play a key role in the regulation of cognitive behavior such as attention, mood, and arousal [82]. These neurons also play role in the development of the brain, mainly the neocortex [83]. The degeneration of LC-neurons has been identified in patients with PD and AD [84]. Also, the classical "monoamine hypothesis of depression" says that the deficiency of NE is a culprit for the cognitive impairment [85]. NE/noradrenaline, the primary neurotransmitter released by the LC neurons targets the adrenergic receptors present on the microglia and astrocytes in the brain [86]. NE-activated ARs on glial cells stimulate the second messenger system and maintain the homeostasis in the brain. Activation of AR on glial cells elicits anti-inflammatory actions, inhibits neuroinflammation, and thereby limits the degeneration of neurons [87]. Moreover, drugs that stimulate the release of NE/NA have potential to reduced inflammation and amyloid pathology in a mouse model of AD [88]. According to Braak's hypothesis, early stage of progression starts in LC before it spreads to SN [89]. Overall, these and many other studies suggest the role of the adrenergic signaling in neurodegeneration. Therefore, enhancing NE/NA signaling, transplanting noradrenergic neurons, or use of drugs that mimic the activity of NA/NE on glial cells have great potential to reverse or halt the progressive degeneration of neurons [90]. The endogenous agonist/ligand for β2-AR is norepinephrine which acts as a neurotrophic factor and can influence protein/DNA synthesis in developing adult brain [91, 92]. NE protects cholinergic and dopaminergic cultured neurons against oxidative stress and catechol moiety of NE plays role in neuroprotection [93, 94]. It suggests that a compound containing catechol moiety, such as β-agonists, can mimic the neuroprotective effects of NE. Treatment with NE stimulates the synthesis of BDNF in astrocytes and neuron *in vitro* and *in vivo* [95, 96] and these neuroprotective effects were reversed by the antagonist of α1, β1, and β2-AR [97].

The use of β2-agonists as an adjunct therapy to L-DOPA in PD was first described in 1994 [98]. Chai et al. showed that the β2-AR activation enhances hippocampal neurogenesis, ameliorates memory deficits, and increases dendritic branching and spine density in a mouse model of Alzheimer's disease [99]. Recently, Mittal et al. have found that β2-AR activation regulates the gene expression of α-synuclein in various animal and *in vitro* models of PD. Salbutamol, a blood-brain-barrier-permeable β2-agonist, reduces expression of SNCA gene via histone-3-lysine-27 acetylation of its promoter and enhancer. They also analyzed the pharmacological history of 4 million Norwegians over 11 years and found that Salbutamol was also associated with reduced risk of developing PD [100]. In a mouse model of Down syndrome, Formoterol, a long-acting β2-AR agonist, causes significant improvement in synaptic density and cognitive functions [101]. Salmeterol (Sal) is an inhaled long-acting highly selective β2-AR agonist which is currently being used as the active ingredient in Advair$^{@}$ as a bronchodilator. Our previous studies and others have shown that Salmeterol has anti-inflammatory and DA-neuroprotective activities, even at very low doses. Pre-treatment with Salmeterol protects DA neurons against LPS- and 1-methyl-4-phenyl-1,2,3,6-tetrahydropyridine (MPTP)-induced toxicity in both *in vitro* and *in vivo* animal models of PD [42, 102]. The mechanism how Salmeterol regulates the activation of microglia is described in **Figure 1**. Collectively, these studies suggest that β2-AR agonists not only protect neurons against degeneration, but also have anti-inflammatory effects, and therefore, hold significant promise for the treatments of a wide variety of neurodegenerative conditions including PD [43]. The clinical efficacy of β2-AR agonists have been examined in various neurological disorders and few of them are summarized in **Table 2**.

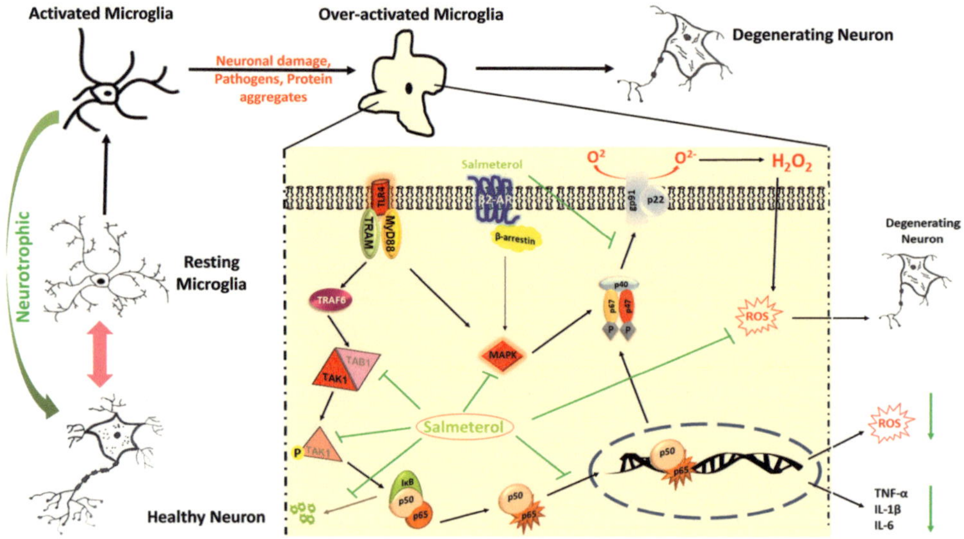

Figure 1.
Schematic of microglia-mediated neurotoxicity and inhibitory effects of Salmeterol on microglial activation.

Disease Condition	Design	Doses	Drug	References
Spinal Cord Injury	Randomized controlled	4mg twice/day for 1st week then 8mg twice/day for 15 weeks	Albuterol	[129]
Alzheimer's Disease	Randomized controlled	20mg/2ml for 12 months	Formoterol	[130]
Multiple Sclerosis	Blinded controlled	4mg/day	Albuterol	[131]
Neuropathic pain	Controlled, double blinded	5mg twice/day for 28 days	Terbutaline	[132]
Memory and Cognition	Randomized controlled	4mg, single oral administration	Salbutamol	[133]
SMA	Uncontrolled	3-8mg/day for 6 months	Albuterol	[134]
ALS	Uncontrolled	60ug/day for 6 months	Clenbuterol	[135]
SBMA	Uncontrolled	20ug/day for 2days, then 40ug/day	Clenbuterol	[136]

***SMA**: Spinal Muscular Atrophy, **SBMA**: Spinal and Bulbar Muscular Atrophy, **ALS**: Amyotrophic Lateral Sclerosis.*

Table 2.
Clinical trials using β2-agonist in neurological conditions.

9. β2-adrenergic receptors and neuroinflammation

Extensive previous investigations into the etiology of PD demonstrate a central role for the inflammatory microglial cell in the progression of PD. Thus, targeting neuroinflammation mediated by microglia may serve as a potential therapeutic benefit in the treatment of PD. Since traditional treatment for PD is aimed only at controlling the disease symptoms, the search for more effective neuroprotective therapies which target the cause of the disease is now receiving significant attention. Studies targeting neuroinflammation are aimed to promote the development of a novel therapeutic approach and aid in the drug discovery for neurodegenerative conditions such as PD.

One such anti-inflammatory approach that has been found to be effective in protection against dopaminergic neurodegeneration is accomplished by natural and therapeutic compounds that activate the β2-AR. Brain cells including neurons, microglia, and astrocytes as well as immune cells express a high density of β2-AR on their surface [66, 103]. Catecholamines such as epinephrine (adrenaline), norepinephrine (noradrenaline), and dopamine are the most abundant catecholamines found in the nervous system. As evidenced by many unrelated studies, catecholamines can modulate the immune response [87, 104]. Further studies have found that the endogenous agonist of β2-AR, norepinephrine (NE), controls microglial motility and functions during pathogenic conditions [105]. NE also protects cortical neurons against microglia-mediated inflammation, while decreased levels of NE enhance microglial activation [106]. One study showed that β2-AR negatively regulates NF-κB activation and stabilizes the NF-κB/IκBα complex via β-arrestin 2 in LPS activated murine macrophages [107]. Interestingly, activation of β2-AR in astrocytes modulates TNF-α-induced inflammatory gene expression *in vitro* and *in vivo*. In addition, an *in vivo* study demonstrated increased expression of β2-AR in glial cells in response to neuronal injury. This suggests that β2-AR may provide a therapeutic target for regulation of glial cell functioning and the inflammatory response in the brain [108]. Activation of β2-AR on astrocytes stimulates the release of trophic factors such as BDNF, bFGF, NGF-1, and TGF-β1 via canonical signaling, showing anti-apoptotic and neuroprotective effects in animal models of cerebral ischemia and excitotoxicity [109, 110]. It has also been shown that noradrenaline acting on β2-AR enhances the expression of anti-inflammatory and neurotrophic cytokine IL-10 in the brain. This suggests an endogenous ligand of β2-AR is neuroprotective during inflammatory conditions in CNS disease pathology [108, 111]. Both canonical and noncanonical signaling of β2-AR can selectively regulate the adaptive immune response [67], since β2-AR are expressed by naïve CD4+ T (T-helper (Th0)) and Th1 cells but absent on Th2 cells [112, 113]. Naïve CD4+ T-cell treated with a β2-AR agonist or NE suppresses the production of interferon (IFN)-γ and IL-2 and affects their differentiation [114]. Collectively, these studies and several others suggest the role of β2-AR in the regulation of immune response.

10. Molecular mechanism of inflammation in PD or molecular mediators of inflammation in PD

10.1 Effect of β2-AR agonists on NF-κB pathway

We have characterized and examined the effects of β2-AR agonists including Salbutamol, Salmeterol, Indacaterol, and Vilanterol on neuroinflammation in models of PD *in vitro* and *in vivo*. However, the short-acting agonists were neuroprotective and able to reduce inflammation *in vitro* at higher doses, but the long-acting agonist showed beneficial effects at low concentration (10^{-9} M) in neurotoxicity and inflammatory models of PD. Salmeterol, a β2-AR agonist, can effectively serve as a therapeutic treatment for PD by inhibiting microglia-mediated inflammatory responses *in vivo*. We have found that Salmeterol functions to inhibit innate pro-inflammatory response in both murine macrophages and microglia through its inhibition of the NF-κB signaling pathways [42]. We have also investigated whether Salmeterol is specific to neuroinflammation in PD or if it can be used as a universal anti-inflammatory drug against other chronic inflammatory diseases. To test this, we used murine macrophages stimulated with LPS from *Porphyromonas gingivalis* (*Pg*LPS), an oral pathogen as an *in vitro* model for the periodontal disease. We have found that Salmeterol shows similar anti-inflammatory effects on *Pg*LPS-stimulated macrophages [115]. Additionally, Feng et al. have also shown neuroprotective effects of β-arrestin2 via endogenous opioid arrest in inflammatory microglial cells [116].

10.2 Effect of β2-AR agonists on MAPK pathway

The agonist-activated β2-AR stimulates MAPK signaling pathway via noncanonical and G-protein independent pathway. Agonist-activated β2-AR reduces phosphorylation of ERK1/2 and p38 MAPK in macrophages stimulated with LPS. In contrast, β2-AR activation stimulates MAPK signaling and TNF-α, IL-12, and NO production in murine macrophages treated with PMA (phorbol 12-myristate-13-acetate) [73]. Similarly, our previous studies have shown that activation of β2-AR with the high concentration of agonist (up to 10^{-5} M) leads to sustained phosphorylation of ERK1/2 and enhanced production inflammatory mediators in murine microglia and macrophages [117]. High-dose treatment of β2-AR agonists on mixed neuroglia culture enhances neurotoxicity via NADPH oxidase activity in the ERK-dependent manner [118]. Like others, we have found that the low-doses of the β2-AR agonist Salmeterol reduces the MAPK activity, NF-κB activation and production of TNF-α in LPS-activated primary microglia [42]. We have also found that low-dose Salmeterol inhibits the phosphorylation of TAK1 (TGF-β-activated kinase1) which is an upstream regulator of NF-κB signaling in LPS-stimulated microglia. We have also found that Salmeterol increases the expression of β-arrestin2 and enhances the interaction between β-arrestin2 and TAB1 (TAK1-binding protein), reduced TAK1/TAB1 mediated activation of NFκB and expression of pro-inflammatory genes. Furthermore, silencing of β-arrestin2 abrogates the anti-inflammatory effects of Salmeterol in LPS-stimulated BV2 cells [119]. These studies suggest that the anti-inflammatory effects of Salmeterol work through the inhibition of pro-inflammatory pathways in microglial cells.

10.3 The β-arrestin-mediated biased effects of β2-AR agonist

Previous findings show that high dose Salmeterol enhances the expression of IL-1β and IL-6 mRNA and protein in unstimulated human monocytes and murine macrophages. These effects were β-arrestin2-dependent but PKA and NF-κB independent, while treatment with ERK1/2 and p38 MAPK inhibitor could reverse this effect [117]. This finding and several others suggest Salmeterol or other long-acting agonist have β-arrestin "biased" signaling of β2-AR. These agonists activate receptors via β-arrestin signaling with a much greater extent than their effect on G-protein-dependent signaling [120]. Our studies suggest that a very low concentration of Salmeterol does not enhance cAMP signaling and its downstream mediators, while it activates the β-arrestin2-mediated signaling events [42]. β-arrestin2 has been shown as a novel regulator of IκB stability via the direct interaction of β-arrestin2 and IκB in HEK293 cells [121]. In addition, β-arrestin2 negatively regulates the activation of NF-κB via direct binding with IκBα [122]. One study showed that overexpression of β-arrestin2 significantly reduces L-DOPA-induced dyskinesia in animal models of PD [123]. Collectively, these studies suggest that β2-AR agonists can be used therapeutically not only to inhibit chronic inflammation and progressive degeneration of neurons, but also to treat some of the most debilitating neurologic symptoms in PD.

10.4 cAMP/PKA/CREB pathway induced by β2-AR

After binding with an agonist or endogenous ligand, β2-AR normally activates the classical cAMP-dependent signaling pathway. The downstream effect of the cAMP/PKA pathway is the phosphorylation and nuclear translocation of the CREB transcription factor which further enhances the expression of cAMP-inducible genes [79]. Activation of CREB via this pathway regulates the synthesis of proteins which are mandatory for neuronal homeostasis [124]. The classical signaling of β2-AR also increases the activity of PGC-1α (Peroxisome proliferator-activated receptor gamma coactivator 1-alpha), which is a key regulator of mitochondrial biogenesis and ROS metabolism [125]. Activation of β2-AR also elevated the release of neurotrophic factors via cAMP/PKA/CREB pathway and provides neuroprotective benefits against degeneration [126]. An endogenous agonist of β2-AR (NE) affects immune cell functions, production of cytokines, and antibody secretion [112]. β2-AR agonists have anti-inflammatory activity and inhibit release of pro-inflammatory mediators via cAMP/PKA/CREB pathway and also by alternate cAMP-dependent pathway (cAMP/Epac1/2) [42, 127, 128]. We have also found that pro-inflammatory effects of high-dose of Salmeterol are through cAMP/Epac pathway, while the anti-inflammatory effects of low-dose of Salmeterol are independent on cAMP and Epac activation [42, 118].

11. From bench to bedside: challenges in translation to the clinic

The β2-AR agonists discussed above are FDA-approved for the treatment of respiratory diseases such as asthma and COPD, but none of these β2-AR agonists are specifically developed for PD. Although, Mittal et al. have found in a Norwegian population that using Salbutamol, a SABA, lower the risk of developing PD whereas the use of Propranolol, a β2-AR antagonist (commonly used to treat hypertension and certain other forms of heart disease) was associated with increased risk of PD [100]. Furthermore, this risk of developing PD was dependent on the duration of Salbutamol intake in those patients. In the patient population who used Salbutamol for at least 6 months, it was expected that 43 would develop PD, but only 23 patients were ultimately diagnosed with the disease (rate ratio 0.66). On the other hand, in the cohort who used Salbutamol for 2 months or less, there was no decreased risk of developing PD in this population [100]. In contrast, patients on Propranolol (which is also used as therapeutic for tremors in PD) for at least 1 year showed a significantly *increased* risk of developing PD compared to patients not on propranolol (rate ratio 2.2). Therefore, it is clear that patients on long-term Salbutamol (a β2-AR agonist) had significantly decreased the risk of developing PD, while patients on long-term propranolol (a β2-AR antagonist) therapy had significantly higher rates of PD, suggesting that β2-AR inhibition is a highly significant risk factor in developing PD. When we compared the effectiveness of Salbutamol to Salmeterol (a more lipophilic drug) in animal models of PD, Salmeterol was much more effective both *in vitro* and *in vivo* in dopaminergic neuroprotection [42]. More importantly, we found that animals given Salmeterol treatment well before the appearance of symptoms in a long-term model of PD showed little evidence of dopaminergic neurodegeneration compared to untreated animals. Taken together, this data suggests that administration of β2-AR agonists may have a profound preventative effect on the development of PD. Since the blood-brain-barrier penetration is a major obstacle in the development of therapeutics targeting CNS disorders, it will be important to consider the importance of lipophilic properties, concentration within the CNS, as well as the specificity, half-life and safety in using β2-AR agonists in older patients before and after the initial appearance of symptoms associated with PD. Consequently, these drugs require further investigation in a large cohort study to assess their utility as a potential therapeutic for PD and other neurodegenerative diseases.

12. Conclusion

Natural or synthetic activation or inhibition of the β2-AR can have profound effects on the development and progression of Parkinson's disease, a chronic neurodegenerative disorder which involves both neuroinflammatory and cellular mechanisms in dopaminergic neurotoxicity. It is now clear that the therapeutic use of β2-AR agonists can both inhibit the cause of neurodegeneration and activate a mechanism that can enhance recovery of patients with this disease, and serves as an important new therapeutic approach to the treatment of chronic neurodegenerative disorders.

Conflict of interest

Authors declare no "conflict of interest."

Author details

Monika Sharma[1] and Patrick M. Flood[2,3,4*]

1 Department of Medical Microbiology and Immunology, Faculty of Medicine and Dentistry, University of Alberta, Edmonton, Alberta, Canada

2 Department of Dentistry, Neuroscience and Mental Health Institute, Faculty of Medicine and Dentistry, University of Alberta, Edmonton, Alberta, Canada

3 Department of Medical Microbiology and Immunology, Neuroscience and Mental Health Institute, Faculty of Medicine and Dentistry, University of Alberta, Edmonton, Alberta, Canada

4 School of Dentistry, Katz Group Centre for Pharmacy and Health Research, University of Alberta, Edmonton, Alberta, Canada

*Address all correspondence to: pflood@ualberta.ca

References

[1] Gao HM, Hong JS. Why neurodegenerative diseases are progressive: Uncontrolled inflammation drives disease progression. Trends in Immunology. 2008;**29**(8):357-365

[2] Poewe W, Seppi K, Tanner CM, Halliday GM, Brundin P, Volkmann J, et al. Parkinson disease. Nature Reviews Disease Primers. 2017;**3**:17013

[3] Nolan YM, Sullivan AM, Toulouse A. Parkinson's disease in the nuclear age of neuroinflammation. Trends in Molecular Medicine. 2013;**19**(3):187-196

[4] Moustafa AA, Chakravarthy S, Phillips JR, Gupta A, Keri S, Polner B, et al. Motor symptoms in Parkinson's disease: A unified framework. Neuroscience and Biobehavioral Reviews. 2016;**68**:727-740

[5] Parkinson's Disease. PubMed - NCBI [Internet]. Available from: https://www.ncbi.nlm.nih.gov/pubmed/?term=25904081[Cited: August 2, 2018]

[6] Löhle M, Storch A, Reichmann H. Beyond tremor and rigidity: Non-motor features of Parkinson's disease. Journal of Neural Transmission. 2009;**116**(11):1483-1492

[7] Gasser T. Molecular pathogenesis of Parkinson disease: Insights from genetic studies. Expert Reviews in Molecular Medicine. 2009;**11**:e22

[8] Spatola M, Wider C. Genetics of Parkinson's disease: The yield. Parkinsonism & Related Disorders. 2014;**20**(Suppl 1):S35-S38

[9] Chade AR, Kasten M, Tanner CM. Nongenetic causes of Parkinson's disease. Journal of Neural Transmission. Supplementum. 2006;**70**:147-151

[10] Wei Z, Li X, Li X, Liu Q, Cheng Y. Oxidative stress in Parkinson's disease: A systematic review and meta-analysis. Frontiers in Molecular Neuroscience. 2018;**11**:236

[11] Doehner J, Genoud C, Imhof C, Krstic D, Knuesel I. Extrusion of misfolded and aggregated proteins—A protective strategy of aging neurons? The European Journal of Neuroscience. 2012;**35**(12):1938-1950

[12] Blaylock RL. Parkinson's disease: Microglial/macrophage-induced immunoexcitotoxicity as a central mechanism of neurodegeneration. Surgical Neurology International. 2017;**8**:65

[13] Ryan BJ, Hoek S, Fon EA, Wade-Martins R. Mitochondrial dysfunction and mitophagy in Parkinson's: From familial to sporadic disease. Trends in Biochemical Sciences. 2015;**40**(4):200-210

[14] Molteni M, Rossetti C. Neurodegenerative diseases: The immunological perspective. Journal of Neuroimmunology. 2017;**313**:109-115

[15] Lehnardt S. Innate immunity and neuroinflammation in the CNS: The role of microglia in toll-like receptor-mediated neuronal injury. Glia. 2010;**58**(3):253-263

[16] Kim C, Ho DH, Suk JE, You S, Michael S, Kang J, et al. Neuron-released oligomeric α-synuclein is an endogenous agonist of TLR2 for paracrine activation of microglia. Nature Communications. 2013;**4**:1562

[17] Hirsch EC, Hunot S. Neuroinflammation in Parkinson's disease: A target for neuroprotection? Lancet Neurology. 2009;**8**(4):382-397

[18] Funk N, Wieghofer P, Grimm S, Schaefer R, Bühring HJ, Gasser T, et al. Characterization of peripheral hematopoietic stem cells and monocytes in Parkinson's disease. Movement Disorders. 2013;**28**(3):392-395

[19] Ginhoux F, Garel S. The mysterious origins of microglia. Nature Neuroscience. 2018;**21**(7):897-899

[20] Li Q, Barres BA. Microglia and macrophages in brain homeostasis and disease. Nature Reviews. Immunology. 2018;**18**(4):225-242

[21] Glezer I, Simard AR, Rivest S. Neuroprotective role of the innate immune system by microglia. Neuroscience. 2007;**147**(4):867-883

[22] Block ML, Zecca L, Hong JS. Microglia-mediated neurotoxicity: Uncovering the molecular mechanisms. Nature Reviews. Neuroscience. 2007;**8**(1):57-69

[23] McGeer PL, Itagaki S, Boyes BE, McGeer EG. Reactive microglia are positive for HLA-DR in the substantia nigra of Parkinson's and Alzheimer's disease brains. Neurology. 1988;**38**(8):1285-1291

[24] Imamura K, Hishikawa N, Sawada M, Nagatsu T, Yoshida M, Hashizume Y. Distribution of major histocompatibility complex class II-positive microglia and cytokine profile of Parkinson's disease brains. Acta Neuropathologica. 2003;**106**(6):518-526

[25] Koshimori Y, Ko JH, Mizrahi R, Rusjan P, Mabrouk R, Jacobs MF, et al. Imaging striatal microglial activation in patients with Parkinson's disease. PLoS One. 2015;**10**(9):e0138721

[26] Perry VH, Holmes C. Microglial priming in neurodegenerative disease. Nature Reviews. Neurology. 2014;**10**(4):217-224

[27] Wang Q, Liu Y, Zhou J. Neuroinflammation in Parkinson's disease and its potential as therapeutic target. Translational Neurodegeneration. 2015;**4**:19

[28] Tambasco N, Romoli M, Calabresi P. Levodopa in Parkinson's disease: Current status and future developments. Current Neuropharmacology. 2017;**16**(8):1239-1252

[29] Connolly BS, Lang AE. Pharmacological treatment of Parkinson disease: A review. Journal of the American Medical Association. 2014;**311**(16):1670-1683

[30] You H, Mariani LL, Mangone G, de Nailly DLF, Charbonnier-Beaupel F, Corvol JC. Molecular basis of dopamine replacement therapy and its side effects in Parkinson's disease. Cell and Tissue Research. 2018;**373**(1):111-135

[31] Sharma M, Arbabzada N, Flood PM. Inflammation: Role in Parkinson's disease and target for therapy. In: Challenges in Parkinson's Disease. Rijeka, Croatia: InTech; 2016

[32] Castaño A, Herrera AJ, Cano J, Machado A. The degenerative effect of a single intranigral injection of LPS on the dopaminergic system is prevented by dexamethasone, and not mimicked by rh-TNF-alpha, IL-1beta and IFN-gamma. Journal of Neurochemistry. 2002;**81**(1):150-157

[33] Manthripragada AD, Schernhammer ES, Qiu J, Friis S, Wermuth L, Olsen JH, et al. Non-steroidal anti-inflammatory drug use and the risk of Parkinson's disease. Neuroepidemiology. 2011;**36**(3):155-161

[34] Gao X, Chen H, Schwarzschild MA, Ascherio A. Use of ibuprofen and risk of Parkinson disease. Neurology. 2011;**76**(10):863-869

[35] Casper D, Yaparpalvi U, Rempel N, Werner P. Ibuprofen protects dopaminergic neurons against glutamate toxicity in vitro. Neuroscience Letters. 2000;**289**(3):201-204

[36] Qian L, Xu Z, Zhang W, Wilson B, Hong JS, Flood PM. Sinomenine, a natural dextrorotatory morphinan analog, is anti-inflammatory and neuroprotective through inhibition of microglial NADPH oxidase. Journal of Neuroinflammation. 2007;**4**:23

[37] Qian L, Wei SJ, Zhang D, Hu X, Xu Z, Wilson B, et al. Potent anti-inflammatory and neuroprotective effects of TGF-bcta1 arc mediated through the inhibition of ERK and p47phox-Ser345 phosphorylation and translocation in microglia. Journal of Immunology. 2008;**181**(1):660-668

[38] Qian L, Block ML, Wei SJ, Lin CF, Reece J, Pang H, et al. Interleukin-10 protects lipopolysaccharide-induced neurotoxicity in primary midbrain cultures by inhibiting the function of NADPH oxidase. The Journal of Pharmacology and Experimental Therapeutics. 2006;**319**(1):44-52

[39] Zhu Y, Liu Z, Peng YP, Qiu YH. Interleukin-10 inhibits neuroinflammation-mediated apoptosis of ventral mesencephalic neurons via JAK-STAT3 pathway. International Immunopharmacology. 2017;**50**:353-360

[40] Zhang F, Qian L, Flood PM, Shi JS, Hong JS, Gao HM. Inhibition of IkappaB kinase-beta protects dopamine neurons against lipopolysaccharide-induced neurotoxicity. The Journal of Pharmacology and Experimental Therapeutics. 2010;**333**(3):822-833

[41] Qian L, Gao X, Pei Z, Wu X, Block M, Wilson B, et al. NADPH oxidase inhibitor DPI is neuroprotective at femtomolar concentrations through inhibition of microglia over-activation. Parkinsonism & Related Disorders. 2007;**13**(Suppl 3):S316-S320

[42] Qian L, Wu H, Chen SH, Zhang D, Ali SF, Peterson L, et al. β2-adrenergic receptor activation prevents rodent dopaminergic neurotoxicity by inhibiting microglia via a novel signaling pathway. Journal of Immunology. 2011;**186**(7):4443-4454

[43] Peterson L, Ismond KP, Chapman E, Flood P. Potential benefits of therapeutic use of β2-adrenergic receptor agonists in neuroprotection and Parkinsonμs disease. Journal of Immunology Research. 2014;**2014**:103780

[44] Qian L, Tan KS, Wei SJ, Wu HM, Xu Z, Wilson B, et al. Microglia-mediated neurotoxicity is inhibited by morphine through an opioid receptor-independent reduction of NADPH oxidase activity. Journal of Immunology. 2007;**179**(2):1198-1209

[45] Li G, Cui G, Tzeng NS, Wei SJ, Wang T, Block ML, et al. Femtomolar concentrations of dextromethorphan protect mesencephalic dopaminergic neurons from inflammatory damage. The FASEB Journal. 2005;**19**(6):489-496

[46] Ahlquist RP. The adrenergic receptor. Journal of Pharmaceutical Sciences. 1966;**55**(4):359-367

[47] Ahlquist RP. Historical perspective. Classification of adrenoreceptors. Journal of Autonomic Pharmacology. 1980;**1**(1):101-106

[48] Ahlquist RP. Adrenergic receptors: A personal and practical view. Perspectives in Biology and Medicine. 1973;**17**(1):119-122

[49] Berthelsen S, Pettinger WA. A functional basis for classification of alpha-adrenergic receptors. Life Sciences. 1977;**21**(5):595-606

[50] Fraser CM, Venter JC. Monoclonal antibodies to beta-adrenergic receptors: Use in purification and molecular characterization of beta receptors. Proceedings of the National Academy of Sciences of the United States of America. 1980;**77**(12):7034-7038

[51] Emorine LJ, Marullo S, Briend-Sutren MM, Patey G, Tate K, Delavier-Klutchko C, et al. Molecular characterization of the human beta 3-adrenergic receptor. Science. 1989;**245**(4922):1118-1121

[52] Siegel GJ, Agranoff BW, Albers RW, et al. Basic Neurochemistry: Molecular, Cellular and Medical Aspects. 6th ed. Philadelphia, PA: Lippincott-Raven; 1999

[53] Fredriksson R, Lagerström MC, Lundin LG, Schiöth HB. The G-protein-coupled receptors in the human genome form five main families. Phylogenetic analysis, paralogon groups, and fingerprints. Molecular Pharmacology. 2003;**63**(6):1256-1272

[54] Garland SL. Are GPCRs still a source of new targets? Journal of Biomolecular Screening. 2013;**18**(9):947-966

[55] Kobilka BK, MacGregor C, Daniel K, Kobilka TS, Caron MG, Lefkowitz RJ. Functional activity and regulation of human beta 2-adrenergic receptors expressed in Xenopus oocytes. The Journal of Biological Chemistry. 1987;**262**(32):15796-15802

[56] Dixon RA, Kobilka BK, Strader DJ, Benovic JL, Dohlman HG, Frielle T, et al. Cloning of the gene and cDNA for mammalian beta-adrenergic receptor and homology with rhodopsin. Nature. 1986;**321**(6065):75-79

[57] Cherezov V, Rosenbaum DM, Hanson MA, Rasmussen SGF, Thian FS, Kobilka TS, et al. High-resolution crystal structure of an engineered human beta2-adrenergic G protein-coupled receptor. Science. 2007;**318**(5854):1258-1265

[58] Palczewski K, Kumasaka T, Hori T, Behnke CA, Motoshima H, Fox BA, et al. Crystal structure of rhodopsin: A G protein-coupled receptor. Science. 2000;**289**(5480):739-745

[59] Collins S, Lohse MJ, O'Dowd B, Caron MG, Lefkowitz RJ. Structure and regulation of G protein-coupled receptors: The beta 2-adrenergic receptor as a model. Vitamins and Hormones. 1991;**46**:1-39

[60] Rands E, Candelore MR, Cheung AH, Hill WS, Strader CD, Dixon RA. Mutational analysis of beta-adrenergic receptor glycosylation. The Journal of Biological Chemistry. 1990;**265**(18):10759-10764

[61] O'Dowd BF, Hnatowich M, Caron MG, Lefkowitz RJ, Bouvier M. Palmitoylation of the human beta 2-adrenergic receptor. Mutation of Cys341 in the carboxyl tail leads to an uncoupled nonpalmitoylated form of the receptor. The Journal of Biological Chemistry. 1989;**264**(13):7564-7569

[62] Dohlman HG, Caron MG, DeBlasi A, Frielle T, Lefkowitz RJ. Role of extracellular disulfide-bonded cysteines in the ligand binding function of the beta 2-adrenergic receptor. Biochemistry. 1990;**29**(9):2335-2342

[63] Carstairs JR, Nimmo AJ, Barnes PJ. Autoradiographic visualization of beta-adrenoceptor subtypes in human lung. The American Review of Respiratory Disease. 1985;**132**(3):541-547

[64] Simon V, Robin MT, Legrand C, Cohen-Tannoudji J. Endogenous G protein-coupled receptor kinase 6 triggers homologous beta-adrenergic receptor desensitization in primary uterine smooth muscle cells. Endocrinology. 2003;**144**(7):3058-3066

[65] Barnes PJ. Beta-adrenoceptors on smooth muscle, nerves and inflammatory cells. Life Sciences. 1993;**52**(26):2101-2109

[66] Tanaka KF, Kashima H, Suzuki H, Ono K, Sawada M. Existence of functional beta1- and beta2-adrenergic receptors on microglia. Journal of Neuroscience Research. 2002;**70**(2):232-237

[67] Lorton D, Bellinger DL. Molecular mechanisms underlying β-adrenergic receptor-mediated cross-talk between sympathetic neurons and immune cells. International Journal of Molecular Sciences. 2015;**16**(3):5635-5665

[68] Barisione G, Baroffio M, Crimi E, Brusasco V. Beta-adrenergic agonists. Pharmaceuticals. 2010;**3**(4):1016-1044

[69] Shukla AK, Xiao K, Lefkowitz RJ. Emerging paradigms of β-arrestin-dependent seven transmembrane receptor signaling. Trends in Biochemical Sciences. 2011;**36**(9):457-469

[70] Luttrell LM, Gesty-Palmer D. Beyond desensitization: Physiological relevance of arrestin-dependent signaling. Pharmacological Reviews. 2010;**62**(2):305-330

[71] Claing A, Laporte SA, Caron MG, Lefkowitz RJ. Endocytosis of G protein-coupled receptors: Roles of G protein-coupled receptor kinases and beta-arrestin proteins. Progress in Neurobiology. 2002;**66**(2):61-79

[72] Shenoy SK, Drake MT, Nelson CD, Houtz DA, Xiao K, Madabushi S, et al. Beta-arrestin-dependent, G protein-independent ERK1/2 activation by the beta2 adrenergic receptor. The Journal of Biological Chemistry. 2006;**281**(2):1261-1273

[73] Szelenyi J, Selmeczy Z, Brozik A, Medgyesi D, Magocsi M. Dual beta-adrenergic modulation in the immune system: Stimulus-dependent effect of isoproterenol on MAPK activation and inflammatory mediator production in macrophages. Neurochemistry International. 2006;**49**(1):94-103

[74] Chu EK, Drazen JM. Asthma: One hundred years of treatment and onward. American Journal of Respiratory and Critical Care Medicine. 2005;**171**(11):1202-1208

[75] Brittain RT. A comparison of the pharmacology of salbutamol with that of isoprenaline, orciprenaline and trimetoquinol. Postgraduate Medical Journal. 1971;**47**(Suppl):11-16

[76] Billington CK, Penn RB, Hall IP. β2 Agonists. Handbook of Experimental Pharmacology. 2017;**237**:23-40

[77] Hizawa N. Pharmacogenetics of β2-agonists. Allergology International. 2011;**60**(3):239-246

[78] Morgan DJ. Clinical pharmacokinetics of beta-agonists. Clinical Pharmacokinetics. 1990;**18**(4):270-294

[79] Johnson M. Molecular mechanisms of beta(2)-adrenergic receptor function, response, and regulation. The Journal of Allergy and Clinical Immunology. 2006;**117**(1): 18-24; quiz 25

[80] Johnson M. Beta2-adrenoceptors: Mechanisms of action of beta2-agonists. Paediatric Respiratory Reviews. 2001;**2**(1):57-62

[81] Chowdhury BA, Dal Pan G. The FDA and safe use of long-acting beta-agonists in the treatment of asthma. The New England Journal of Medicine. 2010;**362**(13):1169-1171

[82] Schwarz LA, Luo L. Organization of the locus coeruleus-norepinephrine

system. Current Biology. 2015;**25**(21):R1051-R1056

[83] Foote SL, Bloom FE, Aston-Jones G. Nucleus locus ceruleus: New evidence of anatomical and physiological specificity. Physiological Reviews. 1983;**63**(3):844-914

[84] Peterson AC, Li CSR. Noradrenergic dysfunction in Alzheimer's and Parkinson's diseases—An overview of imaging studies. Frontiers in Aging Neuroscience. 2018;**10**:127

[85] Schildkraut JJ. The catecholamine hypothesis of affective disorders: A review of supporting evidence. The American Journal of Psychiatry. 1965;**122**(5):509-522

[86] Zorec R, Parpura V, Verkhratsky A. Preventing neurodegeneration by adrenergic astroglial excitation. The FEBS Journal. 2018;**285**(19):3645-3656

[87] Feinstein DL, Kalinin S, Braun D. Causes, consequences, and cures for neuroinflammation mediated via the locus coeruleus: Noradrenergic signaling system. Journal of Neurochemistry. 2016;**139**(Suppl 2):154-178

[88] Heneka MT, Nadrigny F, Regen T, Martinez-Hernandez A, Dumitrescu-Ozimek L, Terwel D, et al. Locus ceruleus controls Alzheimer's disease pathology by modulating microglial functions through norepinephrine. Proceedings of the National Academy of Sciences of the United States of America. 2010;**107**(13):6058-6063

[89] Braak H, Del Tredici K, Rüb U, de Vos RAI, Jansen Steur ENH, Braak E. Staging of brain pathology related to sporadic Parkinson's disease. Neurobiology of Aging. 2003;**24**(2):197-211

[90] Weinshenker D. Long road to ruin: Noradrenergic dysfunction in neurodegenerative disease. Trends in Neurosciences. 2018;**41**(4):211-223

[91] Lauder JM. Neurotransmitters as growth regulatory signals: Role of receptors and second messengers. Trends in Neurosciences. 1993;**16**(6):233-240

[92] Day JS, O'Neill E, Cawley C, Aretz NK, Kilroy D, Gibney SM, et al. Noradrenaline acting on astrocytic β2-adrenoceptors induces neurite outgrowth in primary cortical neurons. Neuropharmacology. 2014;**77**:234-248

[93] Traver S, Salthun-Lassalle B, Marien M, Hirsch EC, Colpaert F, Michel PP. The neurotransmitter noradrenaline rescues septal cholinergic neurons in culture from degeneration caused by low-level oxidative stress. Molecular Pharmacology. 2005;**67**(6):1882-1891

[94] Troadec JD, Marien M, Darios F, Hartmann A, Ruberg M, Colpaert F, et al. Noradrenaline provides long-term protection to dopaminergic neurons by reducing oxidative stress. Journal of Neurochemistry. 2001;**79**(1):200-210

[95] Musazzi L, Rimland JM, Ieraci A, Racagni G, Domenici E, Popoli M. Pharmacological characterization of BDNF promoters I, II and IV reveals that serotonin and norepinephrine input is sufficient for transcription activation. The International Journal of Neuropsychopharmacology. 2014;**17**(5):779-791

[96] Chen MJ, Nguyen TV, Pike CJ, Russo-Neustadt AA. Norepinephrine induces BDNF and activates the PI-3K and MAPK cascades in embryonic hippocampal neurons. Cellular Signalling. 2007;**19**(1):114-128

[97] Juric DM, Loncar D, Carman-Krzan M. Noradrenergic stimulation of BDNF synthesis in astrocytes: Mediation via alpha1- and beta1/beta2-adrenergic

receptors. Neurochemistry International. 2008;**52**(1-2):297-306

[98] Alexander GM, Schwartzman RJ, Nukes TA, Grothusen JR, Hooker MD. Beta 2-adrenergic agonist as adjunct therapy to levodopa in Parkinson's disease. Neurology. 1994;**44**(8):1511-1513

[99] Chai GS, Wang YY, Yasheng A, Zhao P. Beta 2-adrenergic receptor activation enhances neurogenesis in Alzheimer's disease mice. Neural Regeneration Research. 2016;**11**(10):1617-1624

[100] Mittal S, Bjørnevik K, Im DS, Flierl A, Dong X, Locascio JJ, et al. β2-Adrenoreceptor is a regulator of the α-synuclein gene driving risk of Parkinson's disease. Science. 2017;**357**(6354):891-898

[101] Dang V, Medina B, Das D, Moghadam S, Martin KJ, Lin B, et al. Formoterol, a long-acting β2 adrenergic agonist, improves cognitive function and promotes dendritic complexity in a mouse model of Down syndrome. Biological Psychiatry. 2014;**75**(3):179-188

[102] Abdelmotilib H, West AB. Breathing new life into an old target: Pulmonary disease drugs for Parkinson's disease therapy. Genome Medicine. 2017;**9**(1):88

[103] Cash R, Raisman R, Lanfumey L, Ploska A, Agid Y. Cellular localization of adrenergic receptors in rat and human brain. Brain Research. 1986;**370**(1):127-135

[104] Braun D, Madrigal JLM, Feinstein DL. Noradrenergic regulation of glial activation: Molecular mechanisms and therapeutic implications. Current Neuropharmacology. 2014;**12**(4):342-352

[105] Gyoneva S, Traynelis SF. Norepinephrine modulates the motility of resting and activated microglia via different adrenergic receptors. The Journal of Biological Chemistry. 2013;**288**(21):15291-15302

[106] Madrigal JLM, Feinstein DL, Dello Russo C. Norepinephrine protects cortical neurons against microglial-induced cell death. Journal of Neuroscience Research. 2005;**81**(3):390-396

[107] Kizaki T, Izawa T, Sakurai T, Haga S, Taniguchi N, Tajiri H, et al. Beta2-adrenergic receptor regulates toll-like receptor-4-induced nuclear factor-kappaB activation through beta-arrestin 2. Immunology. 2008;**124**(3):348-356

[108] Laureys G, Gerlo S, Spooren A, Demol F, De Keyser J, Aerts JL. β2-adrenergic agonists modulate TNF-α induced astrocytic inflammatory gene expression and brain inflammatory cell populations. Journal of Neuroinflammation. 2014;**11**:21

[109] Culmsee C, Semkova I, Krieglstein J. NGF mediates the neuroprotective effect of the beta2-adrenoceptor agonist clenbuterol in vitro and in vivo: Evidence from an NGF-antisense study. Neurochemistry International. 1999;**35**(1):47-57

[110] Gleeson LC, Ryan KJ, Griffin EW, Connor TJ, Harkin A. The β2-adrenoceptor agonist clenbuterol elicits neuroprotective, anti-inflammatory and neurotrophic actions in the kainic acid model of excitotoxicity. Brain, Behavior, and Immunity. 2010;**24**(8):1354-1361

[111] McNamee EN, Ryan KM, Griffin EW, González-Reyes RE, Ryan KJ, Harkin A, et al. Noradrenaline acting at central beta-adrenoceptors induces interleukin-10 and suppressor of cytokine signaling-3 expression in rat brain: Implications for neurodegeneration. Brain, Behavior, and Immunity. 2010;**24**(4):660-671

[112] Kin NW, Sanders VM. It takes nerve to tell T and B cells what to do. Journal of Leukocyte Biology. 2006;**79**(6):1093-1104

[113] Kohm AP, Sanders VM. Norepinephrine and beta 2-adrenergic receptor stimulation regulate CD4+ T and B lymphocyte function in vitro and in vivo. Pharmacological Reviews. 2001;**53**(4):487-525

[114] Swanson MA, Lee WT, Sanders VM. IFN-gamma production by Th1 cells generated from naive CD4+ T cells exposed to norepinephrine. Journal of Immunology. 2001;**166**(1):232-240

[115] Sharma M, Patterson L, Chapman E, Flood PM. Salmeterol, a long-acting β2-adrenergic receptor agonist, inhibits macrophage activation by lipopolysaccharide from *Porphyromonas gingivalis*. Journal of Periodontology. 2017;**88**(7):681-692

[116] Feng X, Wu CY, Burton FH, Loh HH, Wei L-N. β-Arrestin protects neurons by mediating endogenous opioid arrest of inflammatory microglia. Cell Death and Differentiation. 2014;**21**(3):397-406

[117] Tan KS, Nackley AG, Satterfield K, Maixner W, Diatchenko L, Flood PM. Beta2 adrenergic receptor activation stimulates pro-inflammatory cytokine production in macrophages via PKA- and NF-kappaB-independent mechanisms. Cellular Signalling. 2007;**19**(2):251-260

[118] Qian L, Hu X, Zhang D, Snyder A, Wu HM, Li Y, et al. Beta2 adrenergic receptor activation induces microglial NADPH oxidase activation and dopaminergic neurotoxicity through an ERK-dependent/protein kinase A-independent pathway. Glia. 2009;**57**(15):1600-1609

[119] Sharma M, Flood PM. β-arrestin2 regulates the anti-inflammatory effects of Salmeterol in lipopolysaccharide-stimulated BV2 cells. Journal of Neuroimmunology. 2018;**325**:10-19

[120] Drake MT, Violin JD, Whalen EJ, Wisler JW, Shenoy SK, Lefkowitz RJ. Beta-arrestin-biased agonism at the beta2-adrenergic receptor. The Journal of Biological Chemistry. 2008;**283**(9):5669-5676

[121] Gao H, Sun Y, Wu Y, Luan B, Wang Y, Qu B, et al. Identification of beta-arrestin2 as a G protein-coupled receptor-stimulated regulator of NF-kappaB pathways. Molecular Cell. 2004;**14**(3):303-317

[122] Witherow DS, Garrison TR, Miller WE, Lefkowitz RJ. Beta-Arrestin inhibits NF-kappaB activity by means of its interaction with the NF-kappaB inhibitor IkappaBalpha. Proceedings of the National Academy of Sciences of the United States of America. 2004;**101**(23):8603-8607

[123] Urs NM, Bido S, Peterson SM, Daigle TL, Bass CE, Gainetdinov RR, et al. Targeting β-arrestin2 in the treatment of L-DOPA-induced dyskinesia in Parkinson's disease. Proceedings of the National Academy of Sciences of the United States of America. 2015;**112**(19):E2517-E2526

[124] Zhou HC, Sun YY, Cai W, He XT, Yi F, Li BM, et al. Activation of β2-adrenoceptor enhances synaptic potentiation and behavioral memory via cAMP-PKA signaling in the medial prefrontal cortex of rats. Learning and Memory (Cold Spring Harbor, N.Y.). 2013;**20**(5):274-284

[125] Peterson YK, Cameron RB, Wills LP, Trager RE, Lindsey CC, Beeson CC, et al. β2-adrenoceptor agonists in the regulation of mitochondrial biogenesis. Bioorganic and Medicinal Chemistry Letters. 2013;**23**(19):5376-5381

[126] Counts SE, Mufson EJ. Noradrenaline activation of neurotrophic pathways protects against neuronal amyloid toxicity. Journal of Neurochemistry. 2010;**113**(3):649-660

[127] Theron AJ, Steel HC, Tintinger GR, Feldman C, Anderson R. Can the anti-inflammatory activities of β2-agonists be harnessed in the clinical setting? Drug Design, Development and Therapy. 2013;7:1387-1398

[128] Carnevale D, De Simone R, Minghetti L. Microglia-neuron interaction in inflammatory and degenerative diseases: Role of cholinergic and noradrenergic systems. CNS & Neurological Disorders: Drug Targets. 2007;**6**:388-397

[129] Schilero GJ, Hobson JC, Singh K, Spungen AM, Bauman WA, Radulovic M. Bronchodilator effects of ipratropium bromide and albuterol sulfate among subjects with tetraplegia. Journal of Spinal Cord Medicine. 2018;**41**(1):42-47

[130] Phillips C, Fahimi A, Das D, Mojabi FS, Ponnusamy R, Salehi A. Noradrenergic system in down syndrome and alzheimer's disease a target for therapy. Current Alzheimer Research. 2016;**13**(1):68-83

[131] Khoury SJ, Healy BC, Kivisäkk P, Viglietta V, Egorova S, Guttmann CRG, et al. A randomized controlled double-masked trial of albuterol add-on therapy in patients with multiple sclerosis. Archives of Neurology. 2010;**67**(9):1055-1061

[132] de Quervain D, Schwabe L, Roozendaal B. Stress, glucocorticoids and memory: Implications for treating fear-related disorders. Nature Reviews Neuroscience. 2017;**18**(1):7-19

[133] Salvat E, Yalcin I, Muller A, Barrot M. A comparison of early and late treatments on allodynia and its chronification in experimental neuropathic pain. Molecular Pain. 2018;**14**:17448069-17749683

[134] Soraru G, Pegoraro E, Spinella P, Turra S, D'Ascenzo C, Baggio L, et al. A pilot trial with clenbuterol in amyotrophic lateral sclerosis. Amyotrophic Lateral Sclerosis : Official Publication of the World Federation of Neurology Research Group on Motor Neuron Diseases. 2006;**7**(4):246-248

[135] Kinali M, Mercuri E, Main M, De Biasia F, Karatza A, Higgins R, et al. Pilot trial of albuterol in spinal muscular atrophy. Neurology. 2002;**59**(4):609-610

[136] Querin G, D'Ascenzo C, Peterle E, Ermani M, Bello L, Melacini P, et al. Pilot trial of clenbuterol in spinal and bulbar muscular atrophy. Neurology. 2013;**80**(23):2095-2098

Section 2

Neuroprotection to the CNS

Chapter 5

Protecting the Aging Retina

Shen Nian and Amy C.Y. Lo

Abstract

Aging retina, notably the aging macula, is prone to develop degenerative diseases, such as age-related macular degeneration (AMD), the leading cause of visual loss in individuals aged 65 or above in developed countries. However, current treatments are very limited. Since degeneration, dysfunction, and death of retinal neurons are demonstrated in the pathogenesis of AMD, neuroprotective strategies could serve as a possible way to treat AMD. In this chapter, we will briefly introduce risk factors, pathophysiology, affected neurons, classification, clinical manifestation, and current treatments of AMD. Finally, neuroprotection in both AMD animal models and patients will be discussed.

Keywords: neuroprotection, degeneration, photoreceptor, age-related macular degeneration, vision loss

1. Introduction

Retina, which forms the innermost layer of the eyeball, is considered as the end organ of the central nervous system. Macula, located in the central and posterior part of retina, possesses the highest concentration of photoreceptors and, therefore, is responsible for central vision and high-resolution visual acuity (VA). The fovea is a tiny pit in the center of the macula and in charge of the central, sharpest vision. Unfortunately, the macula is more prone to experience degenerative changes with age, such as age-related macular degeneration (AMD), leading to visual impairment.

Due to the increase of life expectancy worldwide, the size of aging population will become larger in the coming decades [1]. According to "World Population Prospects: The 2017 Revision" released by the United Nations, the number of persons aged 60 or above will more than double, from 962 million in 2017 to 2.1 billion by 2050. The number of people aged 80 or above is predicted to triple by 2050, from 137 million in 2017 to 425 million by 2050 [2]. In mainland of China, people aged above 65 represented 11.4% of the total population in 2017 [3]. As a consequence, the increasing prevalence of AMD will be foreseen globally in the future.

In 2015, it was estimated that AMD was the fourth leading cause of blindness and the third most common cause of moderate to severe visual impairment globally [4]. The meta-analysis performed by Wong and collaborators has estimated that the number of persons with AMD will increase from 196 million in 2020 to 288 million in 2040 worldwide [5]. A similar trend is observed in the projected number of individuals affected by AMD in China, rising from 31.23 million in 2020 to 55.19 million in 2050 [6]. In addition, the prevalence of late AMD did not show significant difference among Asian, European, and North American studies, whereas the number of individuals with early AMD was more in European and North American studies than in Asian studies [7].

2. Anatomy and function of the retina

The eye is composed of three layers, which are the inner retina layer, middle vascular choroid layer, and outer fibrous sclera layer, respectively. Retina, the innermost layer of the eye, consists of two parts: the inner transparent neurosensory retina and outer pigmented epithelial layer—the retinal pigment epithelium (RPE). There is a potential space between neural retina and RPE, called subretinal space. In the neural retina, the neural cell bodies are situated in three layers (**Figure 1**), including the outer nuclear layer (ONL) occupied with nuclei of photoreceptors; the inner nuclear layer (INL) filled with nuclei of horizontal, bipolar, and most of the amacrine cells; as well as ganglion cell layer (GCL) containing nuclei of retinal ganglion cells and the rest of displaced amacrine cells. Additionally, axons and dendrites of these retinal neurons constitute two synaptic layers: the inner plexiform layer (IPL) and outer plexiform layer (OPL) [8]. The RPE cells form a continuous polarized cell monolayer with its apical surface adjacent to the outer segment apices of the photoreceptors and its basal aspect lying on supportive substrate Bruch's membrane.

Retinal neurons are mainly distributed in three layers: ONL with nuclei of photoreceptors; INL with nuclei of horizontal, bipolar, and most of the amacrine cells; and GCL with nuclei of retinal ganglion cells and the rest of displaced amacrine cells. Additionally, axons and dendrites of these retinal neurons constitute two synaptic layers, including IPL and OPL [8].

Photoreceptors are divided into two types: rods, which are dominated in the peripheral retina and responsible for dim light vision and detecting movement and contrast, and cones, which are dominated in the macula, especially the fovea (only cones), and are responsible for bright light vision and sensing color vision

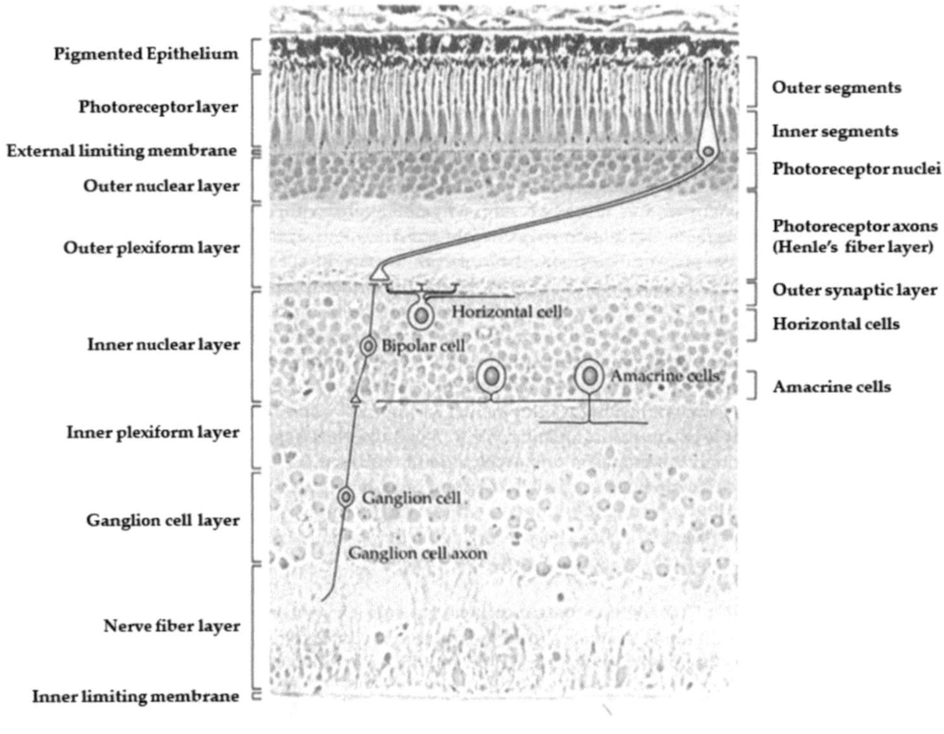

Figure 1.
Retinal layers of the human eye.

and resolution. Outer segments of the photoreceptor are generated by the cell body, and the distal parts of outer segments are phagocytosed by RPE every day. There are abundant mitochondria in inner segments to supply energy for these cells with high metabolic rate. Bruch's membrane, where the RPE cells are tightly attached, consists of five layers including, from inner to outer, the RPE basal lamina, inner collagenous zone, elastic layer, outer collagenous zone, and basement membrane of the choriocapillaris [9].

In brief, the function of the retina is to convert the external light signals into electrical impulses by photoreceptors. The impulses, which are partly integrated locally by horizontal cells and amacrine cells, are then processed by bipolar cells and sent to ganglion cells which further transmit them to the brain. Moreover, RPE is crucial for maintaining the microenvironment of neural retina by exchanging nutrients and wastes between neural retina and choroid, secreting numerous growth factors, phagocytosing shed photoreceptor outer segments (POS), absorbing light and converting all-trans-retinal into 11-cis-retinal [10]. Any defects in any of these functions may result in retinal degeneration, deficits of visual function, and eventually loss of sight.

3. Current knowledge of AMD

3.1 Classification and clinical manifestations

A variety of classification systems are employed for both clinical and basic research of AMD. However, the most commonly used classification is the one defined by the Age-Related Eye Disease Study (AREDS) in terms of the characteristics of drusen (yellow deposits in the macula), abnormal hypo- or hyperpigmentation in the first eye, and how it affected the fellow eye. Based on the above criteria, AMD is classified into 4 categories: (1) no clinical manifestation of AMD if there was no drusen or only non-extensive, small drusen (<63 μm in diameter) in both eyes; (2) mild AMD classified by extensive small drusen, non-extensive intermediate drusen (63–124 μm in diameter), or abnormalities of pigments in at least one eye; (3) intermediate AMD characterized with large drusen (>124 μm in diameter), extensive intermediate drusen, or noncentral geographic atrophy (GA) in at least one eye; (4) advanced AMD defined as central GA or choroidal neovascularization (CNV) resulting in VA less than 20/32 [11–13].

There are two clinical types of advanced AMD: a non-exudative or atrophic (dry) form, accounting for 90% of AMD, and an exudative (wet) form, accounting for only 10% of AMD. The atrophic form is characterized by progressive degeneration of RPE cells and photoreceptors in the macula, affecting central vision to varying degrees over months or years [14, 15]. The exudative form is associated with CNV in the submacular area and subsequent retinal hemorrhage due to leakage of these newly formed fragile blood vessels, leading to severe central vision loss within a very short period of time [15].

At the early stage of AMD, patients are normally asymptomatic and may be diagnosed by the presence of round, yellowish drusen through the routine ophthalmological examinations. Patients with CNV usually suffer from sudden loss of vision, describing unexpected deterioration of central visual field, distorted straight line (metamorphopsia), and/or a dark area in the central visual field (scotoma). On fundus examination, macular edema and hemorrhage are observed and fluorescein angiography shows leakage. In non-exudative AMD, it takes years for patients to develop visual loss gradually, and fundus examination shows a well-defined RPE atrophic area with depigmentation (**Figure 2**) [15].

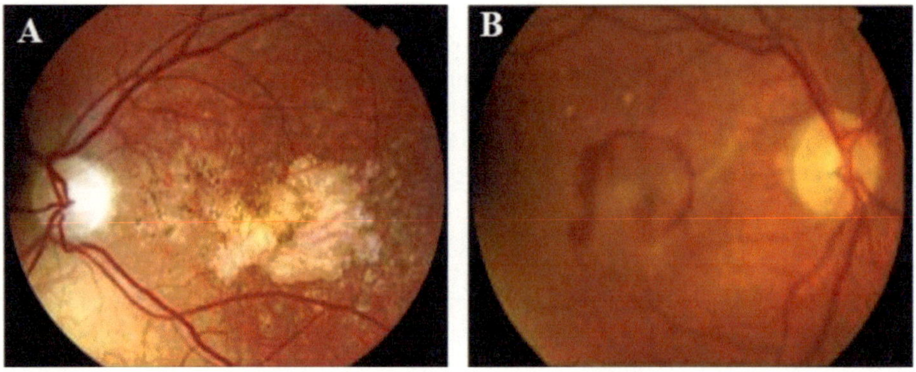

Figure 2.
Representative fundus of AMD patients. (A) Non-exudative form of AMD with a well-defined atrophic region and (B) exudative form of AMD with retinal hemorrhage and macula edema [16].

3.2 Risk factors

A number of risk factors are thought to be related to the pathogenesis of AMD, such as increasing age, smoking, family history of AMD, overweight or obesity, cataract and cataract surgery, eye exposure to the sunlight, etc. Of these factors, age is the strongest risk factor, with around 30% of cases older than 85 years diagnosed with AMD in US [17, 18]. The risk of developing AMD was reported to be more than triple in patients above 75 years than patients aged between 65 and 74 years [18, 19]. In populations of European ancestry, the projected prevalence of late AMD was 0.08% at age 50, 0.33% at age 60, 1.38% at age 70, 5.60% at age 80, and 20.10% at age 90, respectively [20]. Similarly, the estimated prevalence of any AMD increased from 2.44% in persons at the age of 45–49 years to 18.98% in persons at the age of 85–89 years [6].

Ethnicity and family history are also strongly related with AMD prevalence. The highest rate was reported in Caucasians, followed by Hispanics and Asians, while African Americans showed the lowest morbidity rate [21]. Increased risk of AMD was observed in individuals with positive family history, approximately three to six times higher than it was for those from the general population [22]. Furthermore, genetic studies have revealed 34 loci linked to AMD until now, of which complement factor H (CFH) on 1q31.3, age-related maculopathy susceptibility 2 (ARMS2) on 10q26, and CFB/C2 on 6q21.3 were the most discussed genes [15, 23, 24].

The strong relationship between cigarette smoking and development of AMD has been widely accepted [17, 25, 26]. Several studies revealed that current smoking was strongly associated with the progression of AMD [27]. Moreover, past smoking was reported to be a risk factor to develop AMD as well, especially in the rural areas [28]. There was an increased risk for individuals with high intake of several types of fat, such as omega-6 polyunsaturated fatty acids and saturated fats, whereas diets rich in monounsaturated fatty acids might reduce the prevalence of AMD [29]. Inadequate uptake of antioxidants also resulted in AMD, and numerous studies have suggested that antioxidant supplementation, including vitamins, lutein, zeaxanthin, beta-carotene, zinc, etc., helped to halt the progression of AMD to some extent [12, 30, 31].

The effects of sunlight exposure on the occurrence of AMD are controversial. A study performed by Khan and colleagues demonstrated that no significant association between light exposure and AMD was observed when comparing patients at the end stage of AMD with spouse controls [32]. However, other studies suggested that damages induced by either ultraviolet (UV) or visible sunlight might lead to

AMD [33–35]. There was conflicting evidence regarding the role of cataract surgery in the progression of AMD [36]. Earlier, a large population-based research indicated a stronger relationship between cataract surgery with the development of end-stage AMD, especially neovascular AMD, in older patients [36, 37]. On the contrary, several recent clinical studies failed to demonstrate similar results [36, 38].

In addition to the risk factors mentioned above, other relevant factors included hypertension, cardiovascular diseases, iris color, hyperlipidemia, diabetes, alcohol consumption, and so on [22]. According to a systematic review of 18 prospective and cross-sectional studies and 6 case-control studies engaging 113,780 individuals, advancing age, cigarette smoking, previous cataract surgery, and family history of AMD were considered as the strong risk factors of AMD, while hypertension, cardiovascular disease, overweight or obesity, and elevated plasma fibrinogen were regarded as the moderate risk factors [17].

3.3 Pathophysiology

AMD is regarded as a complicated multifactorial disease, although the exact pathogenesis still remains poorly understood. It is generally recognized that impairment of aged RPE cell functions served an important role in the progression of non-exudative AMD. The aged RPE cells were less efficient in phagocytosing and degrading POS, resulting in the progressive accumulation of lipofuscin consisting of phagosomal and lysosomal constituents in the cytoplasm. The cytotoxic elements in lipofuscin, such as bisretinoid fluorophore, were able to generate reactive oxygen species leading to the damage of DNA, lipids, and proteins [39].

Another age-dependent change was the formation of focal extracellular yellow deposits known as drusen, the hallmark of AMD, between the RPE and Bruch's membrane. Accumulation of drusen relied on the changes in the permeability of Bruch's membrane due to the decline of RPE cell functions with aging [40]. In addition, it has been estimated that breakdown of choriocapillaris, which was next to RPE and Bruch's membrane, resulted in insufficient elimination of extracellular wastes causing drusen [41]. According to their shape and size, drusen are separated into small (<125 μm), round hard drusen with well-defined borders, and relatively large, (125–250 μm) soft drusen with poorly defined borders. Soft drusen, especially with depigmentation or pigment abnormalities, was considered to be an indication to give rise to severe vision loss at the late stage.

The mechanism of drusen causing adjacent RPE and photoreceptors' damage was not only dependent on the structural disturbance of RPE and photoreceptor by them, but also on the indirect effects through stimulation of local inflammation and immune system [42, 43]. Components in complement pathway and inflammatory processes were observed in drusen and aged RPE cells that are closely related to drusen [42]. It was demonstrated that in AMD, genetic variation of factor H gene (HF1/CFH) induced the abnormal activities of factor H (the inhibitor of complement cascade) in drusen, leading to the activation of complement pathway and subsequent inflammation in subretinal tissues [44]. Furthermore, in AMD patients, decreased CFH level was reported in smokers than nonsmokers, suggesting the activation of complement cascade might contribute to the significantly higher risk of AMD progression in cases with smoking [45].

DICER1, the ribonucleic acid (RNA)-cleaving enzyme, was decreased in RPE cells in cases with non-exudative AMD. A group of researchers in the University of Kentucky suggested that the lower level of DICER1 led to the lower rate of Alu RNA degradation in RPE cells. Therefore, accumulation of Alu RNA in cytoplasm resulted in RPE cytotoxicity and consequent degeneration via activating NLRP3 inflammasome [46, 47].

In exudative AMD, local nonspecific inflammation stimulated the upregulation of angiogenic factors, such as vascular endothelial growth factor (VEGF), and/or downregulation of anti-angiogenic factors, such as pigment epithelium-derived factor (PEDF), causing CNV development in the avascular outer 1/3 retina. The new vessels are fragile, thus macular edema and hemorrhage occur due to the leakage of these new blood vessels, which finally resulted in the fibrovascular scars in the macula [41, 48].

3.4 Affected neurons

Due to the dysfunction of aged RPE cells, especially the failure of POS phagocytosis, photoreceptor loss occurs subsequently. Although the most remarkable clinical and pathological injuries are present in RPE and its underlying Bruch's membrane, it is the structural and functional disruptions, even the death of photoreceptors, via either atrophic or neovascular process, that take responsibilities for the visual impairment in AMD. Furthermore, the condition of photoreceptors directly reflects the significance of lesions in RPE/Bruch's membrane complex.

It is crucial to identify which type of photoreceptors are most severely damaged in AMD, not only for the potential therapeutic strategies targeting the most affected cells, but also for investigation of mechanism of these pathological changes. The rate of rod and cone degeneration is representative in different situations affecting photoreceptors. In aging retina without age-related maculopathy, the number of cells in the cone-dominated fovea remained stable, while the number of rods in the parafovea was reduced by 30% [49]. In both non-exudative and exudative forms of AMD, photoreceptors were lost. In addition, more rod loss was observed than that of cones; gradually, only degenerated cones were left; finally, all photoreceptors might die [50]. The pathological changes mentioned above were consistent with the functional research exhibiting that scotopic sensitivity decreased more than photopic sensitivity in cases with AMD [51]. Maeda and colleagues revealed the apoptosis of photoreceptors after RPE damage caused by intravitreal injection of ornithine in rats, indicating the important role of RPE cells in maintaining photoreceptor integrity [52]. In AMD patients, apoptotic photoreceptors and RPE cells (TUNEL positive) were observed as well. Most of TUNEL-positive photoreceptors were rods and located at the edge of RPE atrophy. Moreover, Fas was upregulated in apoptotic photoreceptors, indicating Fas/FasL might be involved in the apoptosis process [53]. Studies performed by Kim and collaborators demonstrated that significant reduction of photoreceptors was shown in the areas where RPE cells were totally lost in GA, and where disciform scar formed in wet AMD [54, 55]. In terms of disciform scar, thickness of the scar was closely associated with photoreceptor loss, which means that the thicker the disciform scar, the less photoreceptors survived [56].

In addition to photoreceptors, other retinal neurons are also affected. Joshua et al. first reported that TUNEL-positive cells were detected in the inner side of INL, indicating that these cells might be amacrine cells [53]. In cases with GA, retinal ganglion cells were significantly decreased by 30.7% compared to age-matched control. However, cell nuclei in INL were not significantly different [54]. In wet form of AMD, a decrease of ganglion cells and increase of cells in INL were observed, but there was no significant difference [55].

3.5 Current treatments

Several treatments have been adopted for AMD management. However, current therapeutics can only slow the progression of the disease, trying to delay the onset of vision loss as much as possible.

To date, there are no approved drugs for the dry form of AMD. Therefore, much effort has been made to reduce risk factors. Among various risk factors, oxidative stress induced by inflammation, light exposure, and so on is considered as one of the most important risk factors for the occurrence and progression of AMD. Numerous studies have been conducted to evaluate the relationship between antioxidant nutrition supplementation and AMD. In the Age-Related Eye Disease Study (AREDS), it was demonstrated that oral dietary supplementation containing vitamin C (500 mg/day) and E (400 IU/day), zinc oxide (80 mg/day), cupric oxide (2 mg/day), and beta-carotene (15 mg/day) can decrease the risk of development of AMD from intermediate stage to advanced stage [11]. Since beta-carotene was observed to increase the risk of lung cancer in cigarette smokers, and 80 mg/day zinc is out of tolerance for individuals, there were elimination of beta-carotene, decreased dose of zinc (25 mg/day), and adding of lutein and zeaxanthin in the AREDS2 formula [12, 57, 58]. A 10% reduction in developing to advanced AMD was present in patients treated with AREDS2 formula containing lutein and zeaxanthin [59]. In addition, healthy diet rich in fish, green leafy vegetables, and nuts together with healthy lifestyle are strongly recommended to reduce AMD risk factors [15].

For the wet form of AMD, therapies mainly focus on halting the progression of CNV. Thermal laser photocoagulation is the first treatment to stop the progression of CNV successfully, but with no significant vision improvement and high recurrence of CNV. Photodynamic therapy with verteporfin can selectively damage the CNV tissue without additional injuries of neighboring tissue, but this therapy has no effects on visual improvement either. In terms of upregulation of VEGF in the development of CNV, intravitreal injection of anti-VEGF drugs has been widely used by ophthalmologists as a standard treatment. Anti-VEGF drugs (Pegaptanib sodium, Ranibizumab, and Bevacizumab) for exudative AMD have demonstrated exciting results: the vision in the majority of patients remained stable for 1 year, of which 40% of patients had visual improvement. Surgical intervention to remove CNV and submacular hemorrhage did not improve VA, which is the result of recurrence of CNV [15, 60].

4. Neuroprotection in AMD experimental studies and clinical approaches

Neuroprotection comprises a large number of therapeutic interventions to improve survival of neurons by modifying the structure and function of neurons, and/or their microenvironment. Initially, neuroprotective therapies are focused on central nervous system diseases including stroke, Alzheimer's disease, Parkinson's disease, etc. Since retina is regarded as the end part of central nervous system, a series of neuroprotective strategies has been applied to prevent vision loss by protecting retinal neurons. Furthermore, tremendous neuroprotective strategies are under investigation in both experimental and clinical research.

4.1 Studies in animal models

Neurotrophic factors, belonging to the family of growth factors, have the ability to promote survival of retinal neurons. Ciliary neurotrophic factor (CNTF) is one of the most extensively studied neurotrophic factors for neural retina protection. La Vail and collaborators first reported that intraocular injection of CNTF obviously prevented photoreceptor death from light-induced damage in Sprague Dawley rats [61]. Subsequently, intravitreal injection of adenoviral vector containing CNTF cDNA in rd1 mice, a naturally occurring mouse model for retinal degenerative diseases, demonstrated the reduction of photoreceptor loss, conservation of ONL thickness, and increase of photoreceptor segments' length. Moreover, the amplitudes of a-wave and b-wave in electroretinogram (ERG) were significantly increased compared with those of the control group, suggesting the preservation of retinal functions [62, 63]. Later, long-term protective effects of photoreceptors were shown using adeno-associated virus to deliver CNTF to the retina [64]. In order to sustainably deliver neurotrophic factors, encapsulated human RPE cells secreting CNTF were transplanted into the vitreous of rcd1 dog (a dog model of retinal degeneration). A significant increase of ONL thickness was observed in the treated eye as a result of continuous release of CNTF at the nanogram level [65]. No adverse effects were exhibited in the retina of transplanted eyes during the whole experiment period (7 weeks). In addition, CNTF was proved to protect loss of cone outer segments, an early sign of cone degeneration, indicating that CNTF could not only slow or halt progression of degeneration but also might reverse degeneration [66].

Placental growth factor (PIGF), one of the members of vascular endothelial growth factor family, was believed to prevent neuronal injury in the brain. In the retina, the role of PIGF was exhibited quite differently in *in vitro* and *in vivo* studies. Blue light-induced murine photoreceptor cell death was significantly attenuated after the treatment of PIGF by suppressing caspase-3/7 activity through the mitogen-activated protein kinase (MEK) and phosphoinositide 3-kinase (PI3K) pathway. Anti-PIGF antibody eliminated these protective effects [67]. However, in the light-induced retina-damaged mouse model, PIGF induced decreased ONL thickness and dysfunction of retina. Anti-PIGF antibody diminished neuroretinal injury and disruption of RPE cell-cell junctions after exposure to the white light for 3 h [68]. The opposite effects of PIGF and its antibody in mice were later found to be due to the hyperpermeability of RPE induced by PIGF, leading to the breakdown of retina-blood barrier and subsequent damages [68].

Ursodeoxycholic acid (UDCA) and its taurine-conjugated derivative tauroursodeoxycholic acid (TUDCA) were first found in the bile acid of hibernating bears. They have been used for liver detoxification, dissolution of gallstone and kidney stone, suppression of convulsions, and visual improvement in traditional Chinese medicine for a very long time. According to the theory of modern medicine, UDCA and TUDCA exhibit neuroprotective effects through prevention of cell apoptosis [69]. TUDCA treatment significantly preserved the number and structure of photoreceptors and retinal functions in different murine models of photoreceptor degeneration, including rd10 mice, rd1 mice, BALB/c mice, Bardet-Biedl syndrome type 1 mice, and transgenic P23H rats [69]. Furthermore, TUDCA manifested greater protective effects in cones [70]. *In vitro* studies using photoreceptor 661 W cells revealed that reduced endoplasmic reticulum (ER) stress and improved trafficking of cyclic nucleotide-gated channels in cones contributed to neuroprotective effects of TUDCA [71]. In addition, TUDCA improved phagocytosis of POS in H_2O_2-treated RPE cells via activating Mer tyrosine kinase receptor (MerTK), which indirectly protected photoreceptors [72].

Endogenous and exogenous progesterone have been certified to have neuroprotective effects in brain and retina for several decades. A broad range of studies have been conducted in either light-damaged or genetic murine models of retinal degeneration, demonstrating improvement in photoreceptor survival, decreased gliosis, and reduced retinal dysfunction after administration of progesterone or synthetic progestins [69]. A group of researchers in Spain revealed that rd1 mice orally administered with progesterone (100 mg/kg body weight) at postnatal day 7 (P7) exhibited significantly decreased number of apoptotic cells in ONL in the far peripheral retina and increased amplitude of ERG b-wave at P15, but no significant change was observed at P17. There was also a transient reduced gliosis in the treated rd1 mice [73]. Similar results were observed with oral administration of synthetic progestin, known as the FDA-approved Norgestrel, showing reduction of photoreceptor death by 70 and 75% in light-damaged mouse model and rd10 mice, respectively [74]. The rescue effects were achieved by increasing production of basic fibroblast growth factor and its downstream pro-survival reactive oxygen species [74, 75].

Crystallins, critical family members of small heat shock proteins, have been identified to have novel functions in both retina and RPE as in the lens, such as anti-apoptosis and anti-inflammation. αB-Crystallin is secreted in exosomes released from apical surface of polarized RPE cells and accumulates in the matrix among photoreceptors, and therefore may protect neighboring RPE and photoreceptors [76]. RPE cells in αB-crystallin knockout mice exhibited high susceptibility to oxidative stress and endoplasmic reticulum stress compared to the RPE cells from wild-type mice. Furthermore, RPE cells overexpressing αB-crystallin were more resistant to apoptosis, indicating the protective effects of αB-crystallin [77]. In the mouse model of AMD induced by sodium iodate (NaIO3), absence of αB-crystallin accelerated RPE apoptosis with subsequent death of photoreceptors through upregulation of AKT phosphorylation and expression of peroxisome proliferator-activator receptor-γ, suggesting αB-crystallin, especially the small peptide called mini cry, may play an important role in the protection of retinal degeneration [77]. In order to prolong the life of mini cry in the vitreous, free mini cry was fused to form an elastin-like polypeptide (ELP), which could be detected in the vitreous for up to 2 weeks. One intravitreal injection of ELP-linked peptide prevented RPE cells from apoptosis, inhibited activation of caspase-3 activation, and protected neural retina for up to 1 month after NaIO3 challenge [78].

NF-E2-related factor 2 (Nrf2) is a transcription factor that regulates antioxidant responses in many tissues and cell types, providing protection against oxidative stress. Under the oxidative stress, Nrf2 is translocated from cytoplasm to nucleus, and subsequently binds to the corresponding sites to activate transcription of a wide range of antioxidant genes. In the central nervous system, Nrf2 was proved to slow the neurodegeneration by means of antioxidative stress and neuroinflammation [79]. In the mice undergoing optic nerve crush (an animal model of glaucoma), retinal ganglion cells were significantly decreased than in the wild-type mice. With the treatment of Nrf2 activator, retinal ganglion cell loss was decreased by upregulating gene expression of antioxidant and phase II detoxifying enzymes [80]. After retinal ischemia-reperfusion injury, Nrf2 knockout mice showed greater loss of retinal ganglion cells when compared with the wild-type mice. Moreover, after ischemia-reperfusion injury, Nrf2 activator increased survival of retinal ganglion cells in wild-type mice, but not Nrf2 knockout mice, indicating the neuroprotective effects of Nrf2 [81]. An *in vitro* study has demonstrated that siRNA knockdown of Nrf2 led to significant increase of reactive oxygen species and cell death after blue light exposure in murine photoreceptor cells, suggesting that Nrf2 could be used to protect photoreceptors in AMD and other retinal degeneration [82].

There is no direct evidence that dopamine protects retinal neurons through anti-inflammation, apoptosis, and oxidative stress. However, decreased expression of dopamine may play a negative role in neuron survival in the brain [83]. Thus, dopamine receptor agonists may achieve the goal to promote retinal neuron survival and function by restoring the lost dopamine resulting from certain diseases. Dopamine receptor agonist pramipexole was shown to ameliorate structural and functional injuries in the light-damaged mice, exhibiting decreased photoreceptor death, damage of photoreceptor outer and inner segments, TUNEL-positive cells in ONL, and preservation of a-wave and b-wave in ERG [84]. Additionally, pramipexole inhibited ARPE-19 cell (an immortalized RPE cell line) death after H_2O_2 treatment, suggesting its antioxidative effects [84].

Regular general physical exercise is not only a part of healthy life style, but also a rehabilitation strategy showing neuroprotective effects in numerous diseases. Recently, exercise has been proven to be neuroprotective in animal models of retinal degeneration [85–87]. Wild-type BALB/c mice were forced to exercise for 5 days/week for 2 weeks before being exposed to bright light. Exercised mice showed greater improved amplitude of ERG b-wave and photoreceptor nuclei than the mice without exercise after light exposure [85]. In order to exclude the impact of stress caused by forced exercise, voluntary wheel running was adopted in rd10 mice, which also demonstrated protection of VA, and the number of cones and total photoreceptors [86]. No matter if it was involuntary or voluntary exercise, either one benefited damaged photoreceptors in animal models of retinal degeneration through increased expression of brain-derived neurotrophic factor (BDNF) and activation of tropomyosin-related kinase B (TrkB) signaling pathway [85, 86].

Electrical neurostimulation has developed rapidly in recent years, covering a range of neurological diseases, such as neurostimulation for epilepsy, spinal cord stimulation for chronic pain, brain stimulation for Parkinson's disease, and so on [88]. Electrical neurostimulation in vision research has also made a great improvement, from transcorneal, subretinal, to whole-eye electrical stimulation. Transcorneal electrical stimulation (TES) was performed in SD rats before or after exposure to the intense light for 14 days. Both the stimulation before and after light exposure slowed the progression of photoreceptor degeneration. Furthermore, TES after light exposure exhibited a longer and better protective effect. The neuroprotection effects may result from anti-apoptosis (upregulation of Bcl-2 and downregulation of Bax) and increased expression of neurotrophic factors (CNTF and BDNF) [89]. Subretinal electrical stimulation (SES) in the eye of Royal College of Surgeons (RCS) rats, a commonly used model of retinal degeneration, significantly preserved amplitudes of b-wave and oscillatory potential, and implicit times of a-wave and b-wave, suggesting the preservation of not only photoreceptors, but also signal transmission in the retina [90]. Although the whole-eye electrical stimulation did not provide protection for rod structure in P23H rats, b-wave amplitudes and rod sensitivity were significantly increased [91].

Although the neuroprotective strategies mentioned above have acquired promising results in animal models, damaged human retina can be only partially rescued. Miller and colleagues have investigated the underlying reasons using many animal models. There are three types of cell death including caspase-mediated apoptosis, autophagy-mediated cell death, and necrosis which is regulated by receptor-interacting protein kinases (RIPK). In the animal model of retinal detachment, elevated expression and phosphorylation of RIPK together with activation of caspases were observed. When either RIPK or caspases were inhibited, no obvious rescue effects were demonstrated. However, inhibition of both RIPK and caspases resulted in significant protective effects. Similar results were found in the animal model of retinal degeneration which experienced both photoreceptor and RPE cell death. In this animal model, the dominant type of cell death in photoreceptors was apoptosis, while necrotic RPE cells were mainly exhibited. Therefore, therapies that block both RIPK and caspase pathways may provide more satisfactory neuroprotection [92].

4.2 Current medicine/neuroprotective agents in clinical trials

Neurotech Pharmaceuticals developed an intraocular drug delivery system using encapsulated cell technology, called NT-501, to consistently release CNTF in the vitreous for more than 2 years [93]. A double-masked, randomized, sham-controlled, phase II study enrolling 51 GA patients who were randomly divided into 3 groups, high-dose NT-501 implant, low-dose NT-501 implant, and sham control group, demonstrated promising outcomes after a 2-year evaluation [94]. Another phase II study using intraocular implant secreting CNTF in GA patients has also been completed. In this multicenter, double-masked, randomized, sham-controlled, 1-year, dose-ranging phase II study, VA stabilization, defined as loss of less than 15 letters, was observed in the high-dose group (96.3%) versus low-dose group (83.3%) and sham surgery (75%). All the patients with best corrected VA ≥20/63 in the high-dose group lost less than 15 letters, while it was only 55.6% in the combined group of patients treated with low-dose implant and sham surgery. Additionally, increase of retinal thickness was consistent with the stabilization of VA [95].

Brimonidine, an α-2 agonist, is usually used to treat glaucoma patients in ophthalmology. Since brimonidine has been reported to protect neuroretinal cells in murine models, the effects of this drug on dry AMD patients were under evaluation [96]. A randomized, double-masked, sham-controlled, phase II study involved 119 patients with bilateral GA who were randomly divided into 3 groups including 200 μg treatment, 400 μg treatment, and sham control group. The efficacy and safety of brimonidine on biodegradable implant were evaluated after intravitreal transplantation [97]. But the results were not reliable, so another multicenter study is currently performed with larger samples (311 eyes receiving either treatment of 400 μg brimonidine on biodegradable implant or sham treatment) and longer evaluation period (up to 24 months) [98].

A retrospective analysis was performed to analyze the association between intake of L-DOPA and incidence of AMD. The results showed that the onset of AMD in individuals prescribed with L-DOPA was 8 years later when compared with those without the uptake of L-DOPA. The protective effects of L-DOPA might be through GPR143 (the only known L-DOPA receptor) pathway [99, 100].

There are few studies about exercise as a clinical intervention in AMD, and the published papers are observational research showing the correlation between exercise and prevalence of AMD. A study lasting for 15 years demonstrated that active lifestyle with physical exercise 3 times per week or more was associated with reduced risk to develop wet AMD [101]. In another cohort study, an inverse relation was observed between vigorous exercise (≥3 times/week) and occurrence of intermediate AMD in women, not in men [102]. In addition, low physical exercise was related to the formation of drusen larger than 63 μm [103].

Electrical neurostimulation has already been applied to patients using less invasive approaches, such as transcorneal or whole-eye electrical stimulation, to protect the structure and function of retinal neurons. In a clinical trial conducted by Anastassiou and colleagues, 22 patients with dry AMD received transpalpebral electrical stimulation twice a day for 5 days. Most of the patients demonstrated the improvement in VA and contrast sensitivity at 4 weeks after the treatment; however, only contrast sensitivity was significantly different when compared with sham control group [104]. Similarly, microcurrent stimulation (150 μA for 35 min) once a week for 3 months was applied to both dry and wet AMD. Significantly increased VA was demonstrated in dry AMD, but not in wet AMD. Moreover, the number of patients showing increased VA was twice as those exhibiting deterioration [105].

5. Conclusions

AMD is believed to have stronger relationship with age. With increase of aging population worldwide, more individuals are suffering from visual damage, resulting in poor quality of life for the aged people and elevated cost of medical care. In AMD, it is not only the degeneration of RPE cells, but also the neural retina degeneration, especially photoreceptors, that leads to the visual impairment. Therefore, neuroprotection can be one of the therapeutic strategies to slow, halt, or even reverse the progression of retina degeneration. Although the neuroprotective interventions that are currently investigated in both animal models and patients demonstrate promising results, it is of importance to identify the long-term efficacy and safety of these interventions. Only after that, the therapeutics will be provided to patients to help them maintain vision and further improve the quality of life.

Acknowledgements

This work is supported by the Project of Dominant Discipline Construction in Universities of Shaanxi Province-Basic Medicine, Shaanxi Province, China, awarded to Xi'an Medical University as well as Health and Medical Research Fund, Hong Kong (04150746), awarded to Amy C.Y. Lo.

Conflict of interest

The authors declare no conflict of interest.

Author details

Shen Nian[1] and Amy C.Y. Lo[2]*

1 Department of Pathology, Xi'an Medical University, Xi'an, Shaanxi Province, China

2 Department of Ophthalmology, LKS Faculty of Medicine, The University of Hong Kong, Hong Kong

*Address all correspondence to: amylo@hku.hk

References

[1] Zimmer Z, Jagger C, Chiu CT, Ofstedal MB, Rojo F, Saito Y. Spirituality, religiosity, aging and health in global perspective: A review. SSM—Population Health. 2016;**2**:373-381. DOI: 10.1016/j.ssmph.2016.04.009

[2] United Nations. World Population Prospects: The 2017 Revision. 2017. Available from: https://www.un.org/development/desa/publications/world-population-prospects-the-2017-revision.html [Accessed: September 12, 2018]

[3] National Bureau of Statistics of China. National Population Projections. 2017. Available from: http://data.stats.gov.cn/easyquery.htm?cn=C01&zb=A0306&sj=2016 [Accessed: September 12, 2018]

[4] Bourne RRA, Flaxman SR, Braithwaite T, Cicinelli MV, Das A, Jonas JB, et al. Magnitude, temporal trends, and projections of the global prevalence of blindness and distance and near vision impairment: A systematic review and meta-analysis. The Lancet Global Health. 2017;**5**(9):e888-ee97. DOI: 10.1016/S2214-109X(17)30293-0

[5] Wong WL, Su X, Li X, Cheung CM, Klein R, Cheng CY, et al. Global prevalence of age-related macular degeneration and disease burden projection for 2020 and 2040: A systematic review and meta-analysis. The Lancet Global Health. 2014;**2**(2):e106-e116. DOI: 10.1016/S2214-109X(13)70145-1

[6] Song P, Du Y, Chan KY, Theodoratou E, Rudan I. The national and subnational prevalence and burden of age-related macular degeneration in China. Journal of Global Health. 2017;**7**(2):020703. DOI: 10.7189/jogh.07.020703

[7] Jonas JB, Cheung CMG, Panda-Jonas S. Updates on the epidemiology of age-related macular degeneration. Asia-Pacific Journal of Ophthalmology (Phila). 2017;**6**(6):493-497. DOI: 10.22608/APO.2017251

[8] Oyster CW. Retina I: Photoreceptors and functional organization. In: Oyster CW, editor. The Human Eye: Structure and Function. Sunderland, Mass: Sinauer Associates; 1999. pp. 555-557

[9] Sheridan C, Williams R, Grierson I. Basement membranes and artificial substrates in cell transplantation. Graefe's Archive for Clinical and Experimental Ophthalmology. 2004;**242**(1):68-75. DOI: 10.1007/s00417-003-0800-z

[10] Strauss O. The retinal pigment epithelium in visual function. Physiological Reviews. 2005;**85**(3):845-881. DOI: 10.1152/physrev.00021.2004

[11] Age-Related Eye Disease Study Research Group. Risk factors associated with age-related macular degeneration. A case-control study in the age-related eye disease study: Age-related eye disease study report number 3. Ophthalmology. 2000;**107**(12):2224-2232

[12] Age-Related Eye Disease Study Research Group. A randomized, placebo-controlled, clinical trial of high-dose supplementation with vitamins C and E, beta carotene, and zinc for age-related macular degeneration and vision loss: AREDS report no. 8. Archives of Ophthalmology. 2001;**119**(10): 1417-1436

[13] Bird AC, Bressler NM, Bressler SB, Chisholm IH, Coscas G, Davis MD, et al. An international classification and grading system for age-related maculopathy and age-related macular degeneration. The international ARM

epidemiological study group. Survey of Ophthalmology. 1995;**39**(5):367-374

[14] Sunness JS, Rubin GS, Applegate CA, Bressler NM, Marsh MJ, Hawkins BS, et al. Visual function abnormalities and prognosis in eyes with age-related geographic atrophy of the macula and good visual acuity. Ophthalmology. 1997;**104**(10):1677-1691

[15] Lim LS, Mitchell P, Seddon JM, Holz FG, Wong TY. Age-related macular degeneration. Lancet. 2012;**379**(9827):1728-1738. DOI: 10.1016/S0140-6736(12)60282-7

[16] Arnold J, Sarks S. Age related macular degeneration. Clinical Evidence. 2003;**10**:727-743

[17] Chakravarthy U, Wong TY, Fletcher A, Piault E, Evans C, Zlateva G, et al. Clinical risk factors for age-related macular degeneration: A systematic review and meta-analysis. BMC Ophthalmology. 2010;**10**:31. DOI: 10.1186/1471-2415-10-31

[18] Klein R, Klein BE, Linton KL. Prevalence of age-related maculopathy. The beaver dam eye study. Ophthalmology. 1992;**99**(6):933-943

[19] Leibowitz HM, Krueger DE, Maunder LR, Milton RC, Kini MM, Kahn HA, et al. The Framingham eye study monograph: An ophthalmological and epidemiological study of cataract, glaucoma, diabetic retinopathy, macular degeneration, and visual acuity in a general population of 2631 adults, 1973-1975. Survey of Ophthalmology. 1980;**24**(Suppl):335-610

[20] Rudnicka AR, Jarrar Z, Wormald R, Cook DG, Fletcher A, Owen CG. Age and gender variations in age-related macular degeneration prevalence in populations of European ancestry: A meta-analysis. Ophthalmology. 2012;**119**(3):571-580. DOI: 10.1016/j.ophtha.2011.09.027

[21] Chou R, Dana T, Bougatsos C, Grusing S, Blazina I. Screening for impaired visual acuity in older adults: Updated evidence report and systematic review for the US preventive services task force. Journal of the American Medical Association. 2016;**315**(9):915-933. DOI: 10.1001/jama.2016.0783

[22] Al-Zamil WM, Yassin SA. Recent developments in age-related macular degeneration: A review. Clinical Interventions in Aging. 2017;**12**: 1313-1330. DOI: 10.2147/CIA.S143508

[23] Hughes AE, Orr N, Esfandiary H, Diaz-Torres M, Goodship T, Chakravarthy U. A common CFH haplotype, with deletion of CFHR1 and CFHR3, is associated with lower risk of age-related macular degeneration. Nature Genetics. 2006;**38**(10):1173-1177. DOI: 10.1038/ng1890

[24] Fritsche LG, Igl W, Bailey JN, Grassmann F, Sengupta S, Bragg-Gresham JL, et al. A large genome-wide association study of age-related macular degeneration highlights contributions of rare and common variants. Nature Genetics. 2016;**48**(2):134-143. DOI: 10.1038/ng.3448

[25] Chakravarthy U, Augood C, Bentham GC, de Jong PT, Rahu M, Seland J, et al. Cigarette smoking and age-related macular degeneration in the EUREYE study. Ophthalmology. 2007;**114**(6):1157-1163. DOI: 10.1016/j.ophtha.2006.09.022

[26] Khan JC, Thurlby DA, Shahid H, Clayton DG, Yates JR, Bradley M, et al. Smoking and age related macular degeneration: The number of pack years of cigarette smoking is a major determinant of risk for both geographic atrophy and choroidal neovascularisation. The British Journal of Ophthalmology. 2006;**90**(1):75-80. DOI: 10.1136/bjo.2005.073643

[27] Singh N, Srinivasan S, Muralidharan V, Roy R, Jayprakash V, Raman R. Prevention of age-related macular degeneration. Asia-Pacific Journal of Ophthalmology (Phila). 2017;**6**(6): 520-526. DOI: 10.22608/APO.2017416

[28] Raman R, Pal SS, Ganesan S, Gella L, Vaitheeswaran K, Sharma T. The prevalence and risk factors for age-related macular degeneration in rural-urban India, Sankara Nethralaya rural-urban age-related macular degeneration study, report No. 1. Eye (London, England). 2016;**30**(5):688-697. DOI: 10.1038/eye.2016.14

[29] Parekh N, Voland RP, Moeller SM, Blodi BA, Ritenbaugh C, Chappell RJ, et al. Association between dietary fat intake and age-related macular degeneration in the carotenoids in age-related eye disease study (CAREDS): An ancillary study of the Women's Health Initiative. Archives of Ophthalmology. 2009;**127**(11):1483-1493. DOI: 10.1001/archophthalmol.2009.130

[30] Chew EY, Clemons TE, Agron E, Sperduto RD, Sangiovanni JP, Kurinij N, et al. Long-term effects of vitamins C and E, beta-carotene, and zinc on age-related macular degeneration: AREDS report no. 35. Ophthalmology. 2013;**120**(8):1604-11.e4. DOI: 10.1016/j.ophtha.2013.01.021

[31] Age-Related Eye Disease Study 2 Research Group. Lutein + zeaxanthin and omega-3 fatty acids for age-related macular degeneration: The age-related eye disease study 2 (AREDS2) randomized clinical trial. JAMA. 2013;**309**(19):2005-2015. DOI: 10.1001/jama.2013.4997

[32] Khan JC, Shahid H, Thurlby DA, Bradley M, Clayton DG, Moore AT, et al. Age related macular degeneration and sun exposure, iris colour, and skin sensitivity to sunlight. British Journal of Ophthalmology. 2006;**90**(1):29-32. DOI: 10.1136/bjo.2005.073825

[33] Glazer-Hockstein C, Dunaief JL. Could blue light-blocking lenses decrease the risk of age-related macular degeneration? Retina. 2006;**26**(1):1-4

[34] Margrain TH, Boulton M, Marshall J, Sliney DH. Do blue light filters confer protection against age-related macular degeneration? Progress in Retinal and Eye Research. 2004;**23**(5):523-531. DOI: 10.1016/j.preteyeres.2004.05.001

[35] Liu IY, White L, LaCroix AZ. The association of age-related macular degeneration and lens opacities in the aged. American Journal of Public Health. 1989;**79**(6):765-769

[36] Qian CX, Young LH. The impact of cataract surgery on AMD development and progression. Seminars in Ophthalmology. 2014;**29**(5-6):301-311. DOI: 10.3109/08820538.2014.962166

[37] Wang JJ, Klein R, Smith W, Klein BE, Tomany S, Mitchell P. Cataract surgery and the 5-year incidence of late-stage age-related maculopathy: Pooled findings from the beaver dam and blue mountains eye studies. Ophthalmology. 2003;**110**(10):1960-1967

[38] Rim TH, Lee CS, Lee SC, Kim S, Kim SS. Association between previous cataract surgery and age-related macular degeneration. Seminars in Ophthalmology. 2017;**32**(4):466-473. Epub 2016/04/30

[39] Sparrow JR, Boulton M. RPE lipofuscin and its role in retinal pathobiology. Experimental Eye Research. 2005;**80**(5):595-606. DOI: 10.3109/08820538.2015.1119861

[40] Green WR, McDonnell PJ, Yeo JH. Pathologic features of senile macular degeneration. Ophthalmology. 1985;**92**(5):615-627

[41] Biesemeier A, Taubitz T, Julien S, Yoeruek E, Schraermeyer U. Choriocapillaris breakdown precedes

retinal degeneration in age-related macular degeneration. Neurobiology of Aging. 2014;**35**(11):2562-2573. DOI: 10.1016/j.neurobiolaging.2014.05.003

[42] Johnson LV, Leitner WP, Staples MK, Anderson DH. Complement activation and inflammatory processes in Drusen formation and age related macular degeneration. Experimental Eye Research. 2001;**73**(6):887-896. DOI: 10.1006/exer.2001.1094

[43] Anderson DH, Mullins RF, Hageman GS, Johnson LV. A role for local inflammation in the formation of drusen in the aging eye. American Journal of Ophthalmology. 2002;**134**(3):411-431

[44] Hageman GS, Anderson DH, Johnson LV, Hancox LS, Taiber AJ, Hardisty LI, et al. A common haplotype in the complement regulatory gene factor H (HF1/CFH) predisposes individuals to age-related macular degeneration. Proceedings of the National Academy of Sciences of the United States of America. 2005;**102**(20):7227-7232. DOI: 10.1073/pnas.0501536102

[45] Sepp T, Khan JC, Thurlby DA, Shahid H, Clayton DG, Moore AT, et al. Complement factor H variant Y402H is a major risk determinant for geographic atrophy and choroidal neovascularization in smokers and nonsmokers. Investigative Ophthalmology & Visual Science. 2006;**47**(2):536-540. DOI: 10.1167/iovs.05-1143

[46] Tarallo V, Hirano Y, Gelfand BD, Dridi S, Kerur N, Kim Y, et al. DICER1 loss and Alu RNA induce age-related macular degeneration via the NLRP3 inflammasome and MyD88. Cell. 2012;**149**(4):847-859. DOI: 10.1016/j.cell.2012.03.036

[47] Kaneko H, Dridi S, Tarallo V, Gelfand BD, Fowler BJ, Cho WG, et al. DICER1 deficit induces Alu RNA toxicity in age-related macular degeneration. Nature. 2011;**471**(7338):325-330. DOI: 10.1038/nature09830

[48] Ng EW, Adamis AP. Targeting angiogenesis, the underlying disorder in neovascular age-related macular degeneration. Canadian Journal of Ophthalmology. 2005;**40**(3):352-368. DOI: 10.1016/S0008-4182(05)80078-X

[49] Curcio CA, Millican CL, Allen KA, Kalina RE. Aging of the human photoreceptor mosaic: Evidence for selective vulnerability of rods in central retina. Investigative Ophthalmology & Visual Science. 1993;**34**(12):3278-3296

[50] Curcio CA, Medeiros NE, Millican CL. Photoreceptor loss in age-related macular degeneration. Investigative Ophthalmology & Visual Science. 1996;**37**(7):1236-1249

[51] Curcio CA. Photoreceptor topography in ageing and age-related maculopathy. Eye (London, England). 2001;**15**(Pt 3):376-383. DOI: 10.1038/eye.2001.140

[52] Maeda H, Ogata N, Yi X, Takeuchi M, Ohkuma H, Uyama M. Apoptosis of photoreceptor cells in ornithine-induced retinopathy. Graefe's Archive for Clinical and Experimental Ophthalmology. 1998;**236**(3):207-212

[53] Dunaief JL, Dentchev T, Ying GS, Milam AH. The role of apoptosis in age-related macular degeneration. Archives of Ophthalmology. 2002;**120**(11):1435-1442

[54] Kim SY, Sadda S, Humayun MS, de Juan E Jr, Melia BM, Green WR. Morphometric analysis of the macula in eyes with geographic atrophy due to age-related macular degeneration. Retina. 2002;**22**(4):464-470

[55] Kim SY, Sadda S, Pearlman J, Humayun MS, de Juan E Jr, Melia BM, et al. Morphometric analysis of the macula in eyes with disciform age-related macular degeneration. Retina. 2002;**22**(4):471-477

[56] Green WR, Enger C. Age-related macular degeneration histopathologic studies. The 1992 Lorenz E. Zimmerman lecture. Ophthalmology. 1993;**100**(10):1519-1535

[57] Albanes D, Heinonen OP, Taylor PR, Virtamo J, Edwards BK, Rautalahti M, et al. Alpha-Tocopherol and beta-carotene supplements and lung cancer incidence in the alpha-tocopherol, beta-carotene cancer prevention study: Effects of base-line characteristics and study compliance. Journal of the National Cancer Institute. 1996;**88**(21):1560-1570

[58] Johnson AR, Munoz A, Gottlieb JL, Jarrard DF. High dose zinc increases hospital admissions due to genitourinary complications. The Journal of Urology. 2007;**177**(2):639-643. DOI: 10.1016/j.juro.2006.09.047

[59] Chew EY, Clemons TE, Sangiovanni JP, Danis RP, Ferris FL 3rd, Elman MJ, et al. Secondary analyses of the effects of lutein/zeaxanthin on age-related macular degeneration progression: AREDS2 report No. 3. JAMA Ophthalmology. 2014;**132**(2):142-149. DOI: 10.1001/jamaophthalmol.2013.7376

[60] Abd AJ, Kanwar RK, Kanwar JR. Aged macular degeneration: Current therapeutics for management and promising new drug candidates. Drug Discovery Today. 2017;**22**(11):1671-1679. DOI: 10.1016/j.drudis.2017.07.010

[61] LaVail MM, Unoki K, Yasumura D, Matthes MT, Yancopoulos GD, Steinberg RH. Multiple growth factors, cytokines, and neurotrophins rescue photoreceptors from the damaging effects of constant light. Proceedings of the National Academy of Sciences of the United States of America. 1992;**89**(23):11249-11253

[62] Cayouette M, Behn D, Sendtner M, Lachapelle P, Gravel C. Intraocular gene transfer of ciliary neurotrophic factor prevents death and increases responsiveness of rod photoreceptors in the retinal degeneration slow mouse. The Journal of Neuroscience. 1998;**18**(22):9282-9293

[63] Cayouette M, Gravel C. Adenovirus-mediated gene transfer of ciliary neurotrophic factor can prevent photoreceptor degeneration in the retinal degeneration (rd) mouse. Human Gene Therapy. 1997;**8**(4):423-430. DOI: 10.1089/hum

[64] Schlichtenbrede FC, MacNeil A, Bainbridge JW, Tschernutter M, Thrasher AJ, Smith AJ, et al. Intraocular gene delivery of ciliary neurotrophic factor results in significant loss of retinal function in normal mice and in the Prph2Rd2/Rd2 model of retinal degeneration. Gene Therapy. 2003;**10**(6):523-527. DOI: 10.1038/sj.gt.3301929

[65] Tao W, Wen R, Goddard MB, Sherman SD, O'Rourke PJ, Stabila PF, et al. Encapsulated cell-based delivery of CNTF reduces photoreceptor degeneration in animal models of retinitis pigmentosa. Investigative Ophthalmology & Visual Science. 2002;**43**(10):3292-3298

[66] Li Y, Tao W, Luo L, Huang D, Kauper K, Stabila P, et al. CNTF induces regeneration of cone outer segments in a rat model of retinal degeneration. PLoS One. 2010;**5**(3):e9495. DOI: 10.1371/journal.pone.0009495

[67] Inoue Y, Shimazawa M, Nakamura S, Imamura T, Sugitani S, Tsuruma K, et al. Protective effects of placental growth factor on retinal neuronal

cell damage. Journal of Neuroscience Research. 2014;**92**(3):329-337. DOI: 10.1002/jnr.23316

[68] Izawa H, Inoue Y, Ohno Y, Ojino K, Tsuruma K, Shimazawa M, et al. Protective effects of Antiplacental growth factor antibody against light-induced retinal damage in mice. Investigative Ophthalmology & Visual Science. 2015;**56**(11):6914-6924. DOI: 10.1167/iovs.15-16748

[69] Pardue MT, Allen RS. Neuroprotective strategies for retinal disease. Progress in Retinal and Eye Research. 2018;**65**:50-76. DOI: 10.1016/j.preteyeres.2018.02.002

[70] Oveson BC, Iwase T, Hackett SF, Lee SY, Usui S, Sedlak TW, et al. Constituents of bile, bilirubin and TUDCA, protect against oxidative stress-induced retinal degeneration. Journal of Neurochemistry. 2011;**116**(1):144-153. DOI: 10.1111/j.1471-4159.2010.07092.x

[71] Duricka DL, Brown RL, Varnum MD. Defective trafficking of cone photoreceptor CNG channels induces the unfolded protein response and ER-stress-associated cell death. The Biochemical Journal. 2012;**441**(2):685-696. DOI: 10.1042/BJ20111004

[72] Murase H, Tsuruma K, Shimazawa M, Hara H. TUDCA promotes phagocytosis by retinal pigment epithelium via MerTK activation. Investigative Ophthalmology & Visual Science. 2015;**56**(4):2511-2518. DOI: 10.1167/iovs.14-15962

[73] Sanchez-Vallejo V, Benlloch-Navarro S, Lopez-Pedrajas R, Romero FJ, Miranda M. Neuroprotective actions of progesterone in an in vivo model of retinitis pigmentosa. Pharmacological Research. 2015;**99**:276-288. DOI: 10.1016/j.phrs.2015.06.019

[74] Doonan F, O'Driscoll C, Kenna P, Cotter TG. Enhancing survival of photoreceptor cells in vivo using the synthetic progestin norgestrel. Journal of Neurochemistry. 2011;**118**(5):915-927. DOI: 10.1111/j.1471-4159.2011.07354.x

[75] Ruiz Lopez AM, Roche SL, Wyse Jackson AC, Moloney JN, Byrne AM, Cotter TG. Pro-survival redox signalling in progesterone-mediated retinal neuroprotection. The European Journal of Neuroscience. 2017;**46**(1):1663-1672. DOI: 10.1111/ejn.13604

[76] Kannan R, Sreekumar PG, Hinton DR. Novel roles for alpha-crystallins in retinal function and disease. Progress in Retinal and Eye Research. 2012;**31**(6):576-604. DOI: 10.1016/j.preteyeres.2012.06.001

[77] Zhou P, Kannan R, Spee C, Sreekumar PG, Dou G, Hinton DR. Protection of retina by alphaB crystallin in sodium iodate induced retinal degeneration. PLoS One. 2014;**9**(5):e98275. DOI: 10.1371/journal.pone.0098275

[78] Sreekumar PG, Li Z, Wang W, Spee C, Hinton DR, Kannan R, et al. Intra-vitreal alphaB crystallin fused to elastin-like polypeptide provides neuroprotection in a mouse model of age-related macular degeneration. Journal of Controlled Release. 2018;**283**:94-104. DOI: 10.1016/j.jconrel.2018.05.014

[79] Buendia I, Michalska P, Navarro E, Gameiro I, Egea J, Leon R. Nrf2-ARE pathway: An emerging target against oxidative stress and neuroinflammation in neurodegenerative diseases. Pharmacology & Therapeutics. 2016;**157**:84-104. DOI: 10.1016/j.pharmthera.2015.11.003

[80] Himori N, Yamamoto K, Maruyama K, Ryu M, Taguchi K, Yamamoto M, et al. Critical role of Nrf2 in oxidative stress-induced retinal ganglion cell death. Journal of Neurochemistry. 2013;**127**(5):669-680. DOI: 10.1111/jnc.12325

[81] Xu Z, Cho H, Hartsock MJ, Mitchell KL, Gong J, Wu L, et al. Neuroprotective role of Nrf2 for retinal ganglion cells in ischemia-reperfusion. Journal of Neurochemistry. 2015;**133**(2):233-241. DOI: 10.1111/jnc.13064

[82] Chen WJ, Wu C, Xu Z, Kuse Y, Hara H, Duh EJ. Nrf2 protects photoreceptor cells from photo-oxidative stress induced by blue light. Experimental Eye Research. 2017;**154**:151-158. DOI: 10.1016/j.exer.2016.12.001

[83] Lotharius J, Brundin P. Pathogenesis of Parkinson's disease: Dopamine, vesicles and alpha-synuclein. Nature Reviews Neuroscience. 2002;**3**(12):932-942. DOI: 10.1038/nrn983

[84] Shibagaki K, Okamoto K, Katsuta O, Nakamura M. Beneficial protective effect of pramipexole on light-induced retinal damage in mice. Experimental Eye Research. 2015;**139**:64-72. DOI: 10.1016/j.exer.2015.07.007

[85] Lawson EC, Han MK, Sellers JT, Chrenek MA, Hanif A, Gogniat MA, et al. Aerobic exercise protects retinal function and structure from light-induced retinal degeneration. The Journal of Neuroscience. 2014;**34**(7):2406-2412. DOI: 10.1523/JNEUROSCI.2062-13.2014

[86] Hanif AM, Lawson EC, Prunty M, Gogniat M, Aung MH, Chakraborty R, et al. Neuroprotective effects of voluntary exercise in an inherited retinal degeneration mouse model. Investigative Ophthalmology & Visual Science. 2015;**56**(11):6839-6846. DOI: 10.1167/iovs.15-16792

[87] Chrenek MA, Sellers JT, Lawson EC, Cunha PP, Johnson JL, Girardot PE, et al. Exercise and cyclic light preconditioning protect against light-induced retinal degeneration and evoke similar gene expression patterns. Advances in Experimental Medicine and Biology. 2016;**854**:443-448. DOI: 10.1007/978-3-319-17121-0_59

[88] Edwards CA, Kouzani A, Lee KH, Ross EK. Neurostimulation devices for the treatment of neurologic disorders. Mayo Clinic Proceedings. 2017;**92**(9):1427-1444. DOI: 10.1016/j.mayocp.2017.05.005

[89] Ni YQ, Gan DK, Xu HD, Xu GZ, Da CD. Neuroprotective effect of transcorneal electrical stimulation on light-induced photoreceptor degeneration. Experimental Neurology. 2009;**219**(2):439-452. DOI: 10.1016/j.expneurol.2009.06.016

[90] Ciavatta VT, Mocko JA, Kim MK, Pardue MT. Subretinal electrical stimulation preserves inner retinal function in RCS rat retina. Molecular Vision. 2013;**19**:995-1005

[91] Rahmani S, Bogdanowicz L, Thomas J, Hetling JR. Chronic delivery of low-level exogenous current preserves retinal function in pigmented P23H rat. Vision Research. 2013;**76**:105-113. DOI: 10.1016/j.visres.2012.10.016

[92] Miller JW, Bagheri S, Vavvas DG. Advances in age-related macular degeneration understanding and therapy. US Ophthalmic Review. 2017;**10**(2):119-130. DOI: 10.17925/USOR.2017.10.02.119

[93] U.S. National Library of Medicine, ClinicalTrials.gov. A Study of an Encapsulated Cell Technology (ECT) Implant for Patients with Atrophic Macular Degeneration. 2009. Available from: https://clinicaltrials.gov/ct2/show/NCT00447954 [Accessed: September 23, 2018]

[94] Birch DG, Weleber RG, Duncan JL, Jaffe GJ, Tao W. Randomized trial of ciliary neurotrophic factor delivered by encapsulated cell intraocular implants for retinitis pigmentosa. American Journal of Ophthalmology. 2013;**156**(2):283-92.e1. DOI: 10.1016/j.ajo.2013.03.021

[95] Zhang K, Hopkins JJ, Heier JS, Birch DG, Halperin LS, Albini TA, et al.